KENNETH R. LANG

Sun, Earth and Sky

Springer
Berlin
Heidelberg
New York
Barcelona
Budapest
Hong Kong
London
Milan
Paris
Tokyo

Kenneth R. Lang

SUN, EARTH and SKY

With 133 Figures
Including 61 Color Figures

 Springer

Professor Kenneth R. Lang

Department of Physics and Astronomy
Robinson Hall
Tufts University
Medford, MA 02155, USA

Picture on dust cover: The illustration of the X-ray Sun is from the Yohkoh mission of ISAS, Japan. The X-ray telescope was prepared by the Lockheed Palo Alto Research Laboratory, the National Astronomical Observatory of Japan, and the University of Tokyo with the support of NASA and ISAS.

ISBN 3-540-58778-0 Springer-Verlag Berlin Heidelberg New York

Library of Congress Cataloging-in-Publication Data.
Lang, Kenneth R. Sun, earth and sky / Kenneth R. Lang.
p. cm. Includes bibliographical references and index. ISBN 3-540-58778-0
1. Sun-Popular works. 2. Earth-Popular works. 3. Astronomy-Popular works.
I. Title. QB521.4.L36 1995   523.7–dc20   95-6109

Production editor: C.-D. Bachem, Heidelberg
Cover design: Design Concept, Emil Smejkal, Heidelberg
© for the song "Here comes the Sun" on p. VI: 1969 Harrissongs, Ltd.
Reproductions of the halftone figures: Schneider Repro GmbH, Heidelberg
Data conversion, reproductions of the line drawings, and printing: Appl, Wemding
Binding: J. Schäffer GmbH & Co. KG, Grünstadt

SPIN : 10478158   55/3144 – 543210 – Printed on acid-free paper

Dedicated to
Julia Sarah Lang

Here comes the Sun.
Here comes the Sun and I say
It's alright.
Little darling, it's been a long, cold, lonely winter.
Little darling it feels like years since it's been here.
Here comes the Sun. Here comes the Sun and I say
It's alright.
Little darling, the smiles returning to their faces.

…

Sun, Sun, Sun, Here it comes.

George Harrison

Preface

This book was written for my daughter, Julia; her love and comfort have helped sustain me. It is a pleasure to watch Julia noticing details of everything around her. Children and other curious people perceive worlds that are invisible to most of us.

Here we describe some of these natural unseen worlds, from the hidden heart of the Sun to our transparent air. They have been discovered and explored through the space-age extension of our senses.

A mere half-century ago astronomers were able to view the cosmos only in visible light. Modern technology has now widened the range of our perception to include the invisible realms of subatomic particles, magnetic fields, radio waves, ultraviolet radiation and X-rays. They are broadening and sharpening our vision of the Sun, and providing a more complete description of the Earth's environment. Thus, a marvelous new cosmos is now unfolding and opening up, as new instruments give us the eyes to see the invisible and hands to touch what cannot be felt.

Giant radio telescopes now tune in and listen to the Sun, even on a rainy day. Satellite-borne telescopes, such as the one aboard the Yohkoh, or "sunbeam", satellite, view our daytime star above the absorbing atmosphere, obtaining detailed X-ray images of an unseen Sun. Space probes also directly measure the invisible subatomic particles and magnetic fields in space; for instance, the Ulysses spacecraft has recently sampled the region above the Sun's poles for the first time, and the venerable Voyager space probes may have found the hidden edge of the solar system.

My colleagues have been very generous in providing their favorite pictures taken from ground or space. Numerous diagrams are included, each chosen for the new insight it offers. Every chapter begins with a work of modern art, illustrating how artists depict the mystical and supernatural aspects of the Sun or the subtle variations in illumination caused by changing sunlight. All of these images provide new perspectives on the Sun that warms our soul, and lights and heats our days, and on our marvelous planet Earth that is teeming with an abundance of life.

This book describes a captivating voyage of discovery, recording more than a century of extraordinary accomplishments. Our voyage begins deep inside the Sun where nuclear reactions occur. Here particles of anti-matter, produced during nuclear fusion, collide with their material counterparts, annihilating each other and disappearing in a puff of pure radiative energy.

Neutrinos are also created in the solar core; they pass effortlessly through both the Sun and Earth. Billions of the ghostly neutrinos are passing right through you every second. Massive subterranean neutrino detectors enable us to peer inside the Sun's energy-generating core, but the neutrino count always comes up short. Either the Sun does not shine the way we think it ought to, or our basic understanding of neutrinos is in error. Recent thought suggests that the neutrinos have an identity crisis, transforming themselves into a currently undetectable form.

Today we can peel back the outer layers of the Sun, and glimpse inside by observing its widespread throbbing motions. Surface oscillations caused by sound trapped within the Sun can be deciphered to reveal its internal constitution. This procedure, called helioseismology, has been used to establish the Sun's internal rotation rate; the inside rotates slower than expected, at least in the outer equatorial parts of the Sun.

We then consider the Sun as a magnetic star. Its visible surface is pitted with dark, cool regions, called sunspots, where intense magnetism partially chokes off the outward flow of heat and energy. The sunspots tend to gather together in bipolar groups linked by magnetic loops that shape, mold and constrain the outer atmosphere of the Sun. The number of sunspots changes from a maximum to a minimum and back to a maximum every 11 years or so; most forms of solar activity vary in step with this magnetic sunspot cycle.

As our voyage continues, we discover that the sharp visible edge of the Sun is an illusion; it is enveloped by a tenuous, hot, million-degree gas, called the corona, that is never still. This unseen world is detected at X-ray or radio wavelengths. Such observations show that the apparently serene Sun is continuously changing. It seethes and writhes in tune with the Sun's magnetism, creating an ever-changing invisible realm with no permanent features.

The Yohkoh X-ray telescope has shown that bright, thin magnetized structures, called coronal loops, are in a constant state of agitation, always varying in brightness and structure on all detectable spatial and temporal scales. Dark X-ray regions, called coronal holes, also change in shape and form, like everything else on the restless Sun. A high-speed solar wind squirts out of the holes and rushes past the planets, continuously blowing the Sun away and sweeping interstellar matter aside to form the heliosphere. The heating of the million-degree corona, which expands out to form this solar gale, is one of the great unsolved mysteries of the Sun.

This book next considers violent solar phenomena that are detected at invisible wavelengths and are synchronized with the sunspot cycle of magnetic activity. In minutes, powerful eruptions, called solar flares, release magnetic energy equivalent to billions of nuclear explosions and raise the temperature of Earth-sized regions to tens of millions of degrees. Magnetic bubbles, called coronal mass ejections, expand as they propagate outward from the Sun to rapidly rival it in size; their associated shocks accelerate and propel vast quantities of high-speed particles ahead of them.

Our account then turns toward our home planet, Earth, where the Sun's light and heat permit life to flourish. Robot spacecraft have shown that the space outside our atmosphere is not empty! It is swarming with hot, invisible pieces of the Sun.

The Earth's magnetic field shields us from the eternal solar gale, but the gusty, variable wind buffets our magnetic domain and sometimes penetrates within it. Charged particles that have infiltrated the Earth's magnetic defense can be stored in nearby reservoirs such as the Van Allen radiation belts. Spacecraft have recently released chemicals that can illuminate the space near Earth. Other satellites have found a new radiation belt that contains the ashes of stars other than the Sun.

This voyage continues with a description of the multi-colored auroras, or the northern and southern lights. Solar electrons that apparently enter through the Earth's back door, in the magnetotail, are energized locally within our magnetic realm and are guided into the polar atmosphere where they light it up like a cosmic neon sign.

Unpredictable impulsive eruptions on the Sun produce outbursts of charged particles and energetic radiation that can touch our lives. Intense radiation from a powerful solar flare travels to the Earth in just eight minutes altering its outer atmosphere, disrupting long-distance radio communications, and affecting satellite orbits. Very energetic particles arrive at the Earth within an hour or less; they can endanger unprotected astronauts or destroy satellite electronics. Solar mass ejections travel to our planet in one to four days, resulting in strong geomagnetic storms with accompanying auroras and electrical power blackouts. All of these effects, which are tuned to the rhythm of the Sun's magnetic activity cycle, are of such vital importance that national centers employ space weather forecasters and continuously monitor the Sun from ground and space to warn of threatening solar activity.

The Earth is wrapped in a thin membrane of air that ventilates, protects and incubates us. It acts as a one-way filter, allowing sunlight through to warm the surface but preventing the escape of some of the heat into cold outer space. Without this "natural" greenhouse effect, the oceans would freeze and life as we know it would not exist. Long ago, when the Sun was faint, an enhanced greenhouse effect probably kept the young Earth warm enough to sustain life. Then, as the Sun became more luminous, the terrestrial greenhouse must have been turned down, perhaps by life itself.

Our book next shows how the Sun's steady warmth and brightness are illusory; no portion of the spectrum of the Sun's radiative output is invariant. Recent spacecraft measurements have shown that the Sun's total radiation fades and brightens in step with changing activity levels. It doesn't change by much, only by about 0.1 percent over the 11-year sunspot cycle, but the Sun's invisible ultraviolet and X-ray radiation are up to one hundred times more variable than the visible output. Fluctuations in the Sun's visible and invisible radiation can potentially alter global surface temperatures and influence terrestrial climate and weather, alter the planet's ozone layer, and heat and expand the Earth's upper atmosphere.

To completely assess environmental damage by humans to date, and to fully understand how the environment may respond to further human activity, requires an understanding of solar influences on our planet. We must look beyond and outside the Earth, to the inconstant Sun as an agent of terrestrial change. It can both lessen and compound ozone depletion or global warming by amounts that are now comparable to those produced by atmospheric pollutants.

This book therefore next focuses attention on the Earth's protective ozone layer, that is both modulated from above by the Sun's variable ultraviolet output and threatened from below by man-made chemicals. The ozone layer protects us from the Sun's lethal ultraviolet rays, and progress has been made in outlawing the ozone-destroying chemicals. Our ability to reliably determine the future recovery of the ozone layer will depend on adequate knowledge of how it is damaged or restored by the Sun.

Our voyage then continues with a discussion of the Earth's varying temperature. Large natural fluctuations in the record of global temperature changes mask our ability to clearly detect warming caused by human activity, and numerous complexities limit the certainty of computer models used to forecast future global warming. Strong correlations suggest that the 11-year solar activity cycle may be linked to both the Earth's surface temperature and terrestrial weather.

The "unnatural" greenhouse warming might eventually break through the temperature record, if we keep on dumping waste gases into the air at the present rate. The probable consequences of overheating the Earth as the result of human activity are therefore next examined. They suggest that we should curb the build-up of heat-trapping gases despite the great uncertainties about their current effects.

Yet, it is the Sun that energizes our climate and weather. During the past million years, the climate has been dominated by the recurrent, periodic ice ages, which are mainly explained by changes in the amount and distribution of sunlight on the Earth. Variations in the Earth's orbit and axial tilt slowly alter the distances and angles at which sunlight strikes the Earth, thereby controlling the ponderous ebb and flow of the great continental glaciers.

Smaller, more frequent climate fluctuations are superimposed on the grand swings of the glacial/interglacial cycles; these minor ice ages may result from variations in the activity of the Sun itself. For instance, during the latter half of the seventeenth century the sunspot cycle effectively disappeared; this long period of solar inactivity coincided with unusually cold spells in the Earth's northern hemisphere. Observations of Sun-like variable stars indicate that small, persistent variations in the solar energy output could produce extended periods of global cooling or warming. So, a prudent society will benefit by keeping a close watch on the Sun, the ultimate source of all light and heat on the Earth.

I am grateful to numerous experts and friends who have read individual chapters and commented on their accuracy and completeness. They have greatly improved the manuscript, while also providing encouragement. They include Loren Acton, John Bahcall, Dave Bohlin, Ron Bracewell, Raymond Bradley, Ed Cliver, Nancy Crooker, Brian Dennis, Peter Foukal, Mona Hagyard, David Hathaway, Gary Heckman, Mark Hodor, Bob Howard, Jim Kennedy, Jeff Kuhn, Judith Lean, Bill Livingston, John Mariska, Bill Moomaw, Gene Parker, Art Poland, Peter Sturrock, Einar Tandberg-Hanssen, Jean-Claude Vial, Bill Wagner and Wesley Warren. None of them is responsible for any remaining mistakes in the text!

Special thanks go to my entire family – my wife, Marcella, and my three children Julia, David and Marina. They have had to put up with their crazy father who likes to write books, thereby spending enormous amounts of time that might have been better spent with them.

Medford, February 1995   Kenneth R. Lang

Contents

Wild Geese in Sunlight. Geese flying south in the northern winter, following the Sun's warmth. In this V-shaped pattern of flight, the lead bird deflects currents of air and makes flying easier for those that follow in its wake. The Earth's magnetic field similarly deflects the Sun's wind. (Courtesy of James Tallon)

CHAPTER 1

Good Day, Sunshine

1.1 THE RISING SUN

From earliest times, the Sun has been revered and held in awe. For the Greeks of Aristotle's time, sunlight epitomized the fire in the four basic elements – earth, air, fire and water – from which all things arose. Ancient solar observatories, dedicated to the divine Sun-god Ra, can still be found in Luxor, that enchanting city by the Nile; giant Egyptian obelisks, erected thousands of years ago in Luxor and Heliopolis (City of the Sun), now cast their shadows in sundial fashion across points of Paris, London, and Rome.

According to this incantation from Ptolemaic Egypt:

Opening his two eyes, [Ra, the Sun god] cast light on Egypt, he separated night from day. The gods came forth from his mouth and mankind from his eyes. All things took their birth from him.[1]

In the Old Testament's *Book of Genesis*, we find that the Earth was initially a vast waste, covered by darkness, until God said "Let there be light" and the Sun separated day from night. And since the time of Zarathustra (Greek Zoroastres, seventh century B.C.), we have associated light with good, beauty, truth and wisdom, in sharp contrast with the dark forces of evil. The war between good and evil in the *Dead Sea Scrolls* is, for example, depicted as a battle of the Sons of Light against the Sons of Darkness.

The Mayas, Toltecs and Aztecs of Central America had a host of Sun gods; the Aztecs regularly fed the hearts of sacrificial victims to the Sun to strengthen it on its daily journey. Shintoism, a religion based on Sun-worship, has continued for thousands of years in Japan, the land of the rising Sun. Today you can celebrate sunrise, as I have, with Hindu worshipers on the terraced banks, or ghats, along the Ganges river at Benares, India's holiest city.

Nowadays, fire symbolically lights the darkness in many of our rituals, including the torch of the Olympic games, and candlelight vigils or dimmed lights that bring focus to tragic events and times of crisis. In everyday life, most of us feel happier on bright days than on gloomy ones, so cheerful people have a "sunny" disposition while an unhappy day is a "dark" one. And throughout the world, oiled Sun-worshippers lie

Fig. 1.1. Woman in Morning Sun. 1818. This glowing sunrise seems to have a transcendental, mystical quality; it was painted by Caspar David Friedrich, who once compared the "radiating beams of light" in one of his paintings to "the image of the eternal life-giving Father". (Courtesy of Museum Folkwang, Essen)

on tranquil beaches, letting the summer Sun warm their bodies and give them strength.

Painters have often used the Sun to portray a spiritual relationship with nature, particularly at sunrise or sunset (Fig. 1.1). Modern artists also depict the elusive, varying qualities of the Sun – the source of all light and color.

Sunlight seems to dominate, consume and absorb everything in one of Joseph M. W. Turner's paintings (Fig. 1.2). Examples of the artist's perspective on the Sun are provided at the beginning of every chapter in this book, each chosen for its artistic value and for the new insights it offers.

Here you will find "another light, a stronger Sun" portrayed by Vincent Van Gogh; a powerful, yellow Sun that blazes forth with an almost supernatural radiance. Claude Monet's portrayal of sunrise is included – the one that inaugurated the impressionist movement of painting. He used entire sequences of paintings to depict the subtle changes that varying sunlight causes in our perception of objects, such as haystacks or the cathedral at Rouen.

The beginning chapter pages also include Joan Miró's powerful red disk of the Sun, linked to young women on Earth or to the stars beyond. In other instances, we reproduce works that separate the Sun from any reference to the Earth or sky; they show that the Sun can be an intense source of pleasure and beauty by itself.

Writers have also been captivated by the light of the Sun, from the American author Ralph Waldo Emerson, who wrote that pure light was "the reappearance of the original soul", to the German philosopher Friedrich Nietzsche who wrote in *Thus Spoke Zarathustra* :

Fig. 1.2. Regulus. 1837. In this painting by Joseph M. W. Turner, every object is in a fiery, misty state. Brilliant yellow rays of light come down from a central, all-powerful Sun, absorbing and consuming everything else. The picture is named after the Roman general Regulus who was punished for his betrayal of the Carthaginians by having his eyelids cut off, and being blinded by the glare of the Sun. Regulus, who is apparently absent from the scene, has been identified with the spectator, staring into the blinding Sun. (Courtesy of the Tate Gallery, London)

The Moon's love affair has come to an end!
Just look! There it stands; pale and dejected – before the dawn!
For already it is coming, the glowing Sun –
its love of the Earth is coming!
All Sun-love is innocence and creative desire!
Just look how it comes impatiently over the sea!
Do you not feel the thirst and hot breath of its love?[2]

The Sun warms our soul, and lights and heats our days!

Today's astronomers may describe the Sun, and our dependence upon it, in greater scientific detail than artists or writers, but that in no way diminishes their sense of awe for the life-sustaining, even mystical power of the Sun.

1.2 FIRE OF LIFE

Without the Sun's light and heat, life would quickly vanish from our planet. Sunlight is, for example, absorbed by green plants where it strikes chlorophyll, giving it the energy to break water molecules apart and energize photosynthesis; plants thereby use the Sun's energy to live and grow, giving off oxygen as a byproduct (Fig. 1.3). Animals eat these plants for nourishment, so photosynthesis is ultimately the source of all food. Both the oxygen we breathe and the food we consume rely on the Sun's presence.

The Sun also provides, directly or indirectly, almost all the energy on Earth. In burning coal or oil, we are using energy that came from the Sun in ages past when sunlight was trapped in plants and compressed into fossil fuel. It is the Sun's heat that also powers the winds and cycles water from sea to rain.

Thus, our lives depend on the Sun's continued presence and steady warmth. A small increase in the amount of sunlight we receive could boil away our water and make the Earth too hot to inhabit; a small decrease could freeze the oceans and bring back the great ice ages.

We receive just enough energy from the Sun to keep most of our water liquid, and this seems to have made the difference between the development of life on Earth and its apparent absence elsewhere in the solar system. The surface of Venus, the next planet closest to the Sun, is hot enough to melt lead, so any former oceans on Venus were boiled away long ago. Further away from the Sun, the planet Mars is now frozen into a global ice age; it cannot now rain on Mars, and liquid water cannot exist there.

Fig. 1.3. Sunflowers. The Sun sustains all living creatures and plants on Earth. Here the eyes of sunflowers turn in unison to follow their life-sustaining star. (Courtesy of Charles E. Rodgers)

1.3 SUNLIGHT

Occasionally the mixture of colors in a beam of sunlight is spread before our eyes, as when raindrops act like tiny prisms, bending white sunlight into its separate colors and giving us a rainbow (Fig. 1.4). Our eye and brain translate the visible sunlight into colors. From long to short waves, they correspond to red, orange, yellow, green, blue and violet. Plants appear green, for example, because they absorb red sunlight and reflect the green portion of the Sun's light.

However, we each color our world somewhat differently. There are subtle differences in the exact shade of color we perceive, depending on the molecules in our eye's detection system. So, even people with normal eyesight do not always see eye to eye.

Why does the human eye respond just to visible light? Adaptation. The most intense radiation of the Sun is emitted at these wavelengths, and our atmosphere permits it to reach the ground. If our eyes were not sensitive to visible sunlight, we could not identify objects or move around on the Earth's surface. The sensitivity of our eyes is matched to the tasks of vision.

Fig. 1.4. Light Painting. This picture was made by using crystals to liberate the spectral colors in visible sunlight, refracting them directly onto a photographic plate. It was obtained in the rarefied atmosphere atop Hawaii's Mauna Kea volcano, where many of the world's best telescopes are located. (Courtesy of Eric J. Pittman, Victoria, British Colombia)

Rattlesnakes, on the other hand, have infrared-sensitive eyes that enable them to see the heat radiated by animals at night. (Infrared radiation has wavelengths that are slightly longer than visible red light.) You can similarly locate other people at night by using infrared sensors that detect their heat, and satellites use infrared telescopes to detect the heat radiated by rocket exhaust and by large concentrations of people and vehicles.

1.4 DAYTIME STAR

All stars are suns, kin to our own daytime star. Indeed, the Sun is just one of about one hundred billion stars in our Galaxy, the Milky Way, and countless billions of galaxies stretch out in the seemingly boundless Universe. But the Sun is a special star; it is our only daytime star! As Francis William Bourdillon stated:

> The night has a thousand eyes,
> And the day but one;
> Yet the light of the bright world dies,
> With the dying Sun.[3]

The Sun is a quarter million times closer to us than the next nearest star. Because of this closeness, the Sun is a hundred billion times brighter than any other star. Its brilliance provides ample light for the most exacting studies of its chemical constituents, magnetic fields, and surface oscillations. This blessing can also be a curse, for the Sun's heat can melt mirrors or burn up electronic equipment when focused to high intensity. For this reason, special mirror configurations are used to reduce the concentration of visible sunlight, while still producing large images that contain fine detail (Fig. 1.5).

The Sun's proximity allows a level of detailed examination unique among stars. While most other stars appear only as unresolved spots of light in the best telescopes, the Sun reveals its surface features in exquisite detail. Ground-based optical telescopes can resolve structures on the Sun's visible surface that are about 700 kilometers across, about the distance from Boston to Washington, D. C. and about three-quarters the size of France; that is comparable to seeing the details on a coin from one kilometer away.

Yet, the resolution of ground-based telescopes is limited by turbulence in the Earth's atmosphere; it reduces the clarity of the Sun's image at visible wavelengths. (Similar variations cause the stars to twinkle at night, and produce a loss of resolution when we look across a hot road in a desert.) The best visible images with even finer detail can be obtained from the unique vantage point of outer space, using satellite-borne telescopes unencumbered by the limits of our atmosphere.

Fig. 1.5. Eyes on the Sun. Scattered sunlight colors the McMath solar telescope a stunning red, while stars mark trails across the evening sky (*top*). A moveable heliostat, perched atop this telescope, follows the Sun and directs its light downward through the long fixed shaft of the telescope (*bottom*). A figured mirror at the bottom reflects and focuses the sunlight toward the observation room. The shaft's axis is parallel to the rotation axis of the Earth, and about three fifths of it is underground. It is kept cool by pumping cold water through tubes in the exterior skin, thereby reducing turbulence in the air inside and keeping the Sun's image steady. (Courtesy of William C. Livingston, NOAO)

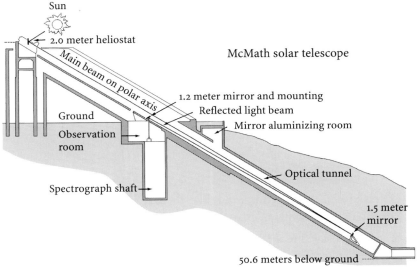

The other stars are so far away that their surfaces remain unresolved with even the largest telescopes. The Sun therefore permits examination of physical phenomena and processes that cannot be seen in detail on other stars. Furthermore, the Sun's basic properties provide benchmarks and boundary conditions for the study of stellar structure and evolution.

So, all astronomers do not work in the dark. Many of them closely scrutinize our daytime star, deciphering some of the most fundamental secrets of nature.

1.5 COSMIC LABORATORY

The Sun can be a site to test physical theories under conditions not readily attainable in terrestrial laboratories. For example, in contrast to our material world, the Sun's core also contains small quantities of short-lived anti-matter. When the two kinds of particles collide, they destroy each other, releasing pure radiative energy. Other subatomic particles made inside the Sun pass effortlessly through both the Sun and the Earth. Recent observations of these ghostlike neutrinos may help us understand the subatomic realm at levels beyond the reach of today's most powerful particle accelerators, perhaps providing new insight to a theory that will unify all the forces of nature.

Our home star also gives us a close-up view of cosmic upheaval and violence. From afar, the Sun seems to be calm, serene, and unchanging, a steadily-shining beacon in the sky; but detailed observations reveal an active, ever-changing Sun. Violent storms and explosive eruptions create gusts in its steady flow of heat and light. The Sun therefore provides us with a unique, high-resolution perspective of the perpetual change and violent activity that characterize much of the Universe.

Thus, we now understand the Sun as a unique star, one so close that it serves as a cosmic laboratory for understanding the physical processes that govern all the other stars, as well as the entire Universe. Everything we learn about the Sun has implications throughout the cosmos, including planet Earth. As examples, observations of the Sun's visible radiation unlocked the chemistry of the Universe, and investigations of the Sun's internal furnace paved the way to nuclear energy.

1.6 INGREDIENTS OF THE SUN

Celestial objects are composed, like the Earth and we ourselves, of individual particles of matter called atoms. But the atoms consist largely of empty space, just as the room you may be sitting in appears mostly empty. A tiny, heavy, positively charged nucleus lies at the heart of an atom, surrounded by a cloud of relatively minute, negatively charged electrons that occupy most of an atom's space and govern its chemical behavior.

The nucleus is itself composed of positively-charged protons and neutral particles, called neutrons; the proton and neutron are both 1840 times heavier than the electron. (Ernest Rutherford established the existence of a small, massive positive nucleus at the center of the atom in 1911; the neutron was discovered by James Chadwick in 1932.) The charge on a proton is exactly equal to that on an electron, so the complete atom, in which the number of electrons equals the number of protons, is electrically neutral.

Fig. 1.6. Visible Solar Spectrum. A spectrograph has spread out the visible portion of the Sun's radiation into its spectral components, displaying radiation intensity as a function of wavelength. When we pass from short wavelengths to longer ones (*bottom* to *top*), the spectrum ranges from violet through blue, green, yellow, orange and red. Dark gaps in the spectrum, called Fraunhofer absorption lines, represent absorption by atoms in the Sun. The wavelengths of these absorption lines can be used to fingerprint the elements in the Sun, and the relative darkness of the lines establish the relative abundance of these elements. (Courtesy of National Solar Observatory/ Sacramento Peak, NOAO)

The simplest and lightest atom consists of a single electron circling around a nucleus composed of a single proton without any neutrons; this is an atom of hydrogen. The nucleus of helium, another abundant light atom, contains two neutrons and two protons, and so has two electrons in orbit.

The electrons orbit the nucleus at relatively large distances. A nucleus is therefore much smaller than the atom, about 40 000 times smaller. To put the components into perspective, if the nucleus of an atom were the size of an orange, an atom would be as large as a mountain.

Radiation is emitted or absorbed by atoms when an electron jumps from one orbit to another, each jump being associated with a specific energy and a single wavelength, like one pure note. If an electron jumps from a low-energy orbit to a high-energy one, it absorbs radiation at this wavelength; radiation is emitted at exactly the same wavelength when the electron jumps the opposite way. This unique wavelength is related to the difference between the two orbital energies.

When a cool, tenuous gas is placed in front of a hot, dense one, atoms in the cool gas absorb radiation at specific wavelengths, thereby producing dark absorption lines. Such features are found superimposed on the colors of sunlight (Fig. 1.6). They are called absorption lines because they look like a dark line when the Sun's radiation intensity is displayed as a function of wavelength; such a display is called a spectrum. The term Fraunhofer absorption lines is also used, recognizing Fraunhofer's 1814 investigation of them. When a tenuous gas stands alone and is heated to incandescence, the energized electrons produce emission lines that shine at precisely the same wavelengths as the dark ones.

The Sun's dark absorption lines and bright emission lines carry messages from inside the atom, and help us determine its internal behavior.

FOCUS 1A
Composition of the Stars

In the mid-nineteenth century, Gustav Kirchhoff discovered a method
for determining the ingredients of the stars. Working with Robert Bunsen, inventor of the Bunsen burner, Kirchhoff showed that every chemical element, when heated to incandescence, emits brightly-colored spectral signatures, or emission lines, whose unique wavelengths coincide
with those of the dark absorption lines in stellar spectra.

By comparing the Sun's absorption lines with emission lines of elements vaporized in the laboratory, Kirchhoff identified in the solar atmosphere several elements known on Earth, including sodium, calcium
and iron. This suggested that stars are composed of terrestrial elements
that are vaporized at the high stellar temperatures, and it unlocked the
chemistry of the Universe. As Bunsen wrote in 1859:

> At the moment I am occupied by an investigation with Kirchhoff
> which does not allow us to sleep. Kirchhoff has made a totally
> unexpected discovery, inasmuch as he has found out the cause for
> the dark lines in the solar spectrum and can produce these lines
> artificially intensified both in the solar spectrum and in the
> continuous spectrum of a flame, their position being identical with
> that of Fraunhofer's lines. Hence the path is opened for the
> determination of the chemical composition of the Sun and the fixed
> stars. [4]

In a brilliant doctoral dissertation, published in 1929, Cecilia Payne
showed that virtually all bright middle-aged stars have the same composition. Miss Payne's calculations also indicated that hydrogen is by far the
most abundant element in the Sun and other stars. But she could not believe that the composition of stars differed so enormously from that of
the Earth, where hydrogen is rarely found, so she mistrusted her understanding of the hydrogen atom. We now know that hydrogen is the most
abundant element in the Universe, and that there was nothing wrong
with her calculations. The Earth just does not have sufficient gravity to
retain hydrogen in its atmosphere; its original hydrogen just evaporated
away while the Earth was forming and has long since escaped.

Subsequent observations have shown that very old stars have practically no elements other than hydrogen and helium; these stars have probably existed since our Galaxy formed. Middle-aged stars like the Sun contain heavier elements. They must have formed from the ashes of previous
generations of stars that have fused lighter elements into heavier ones.

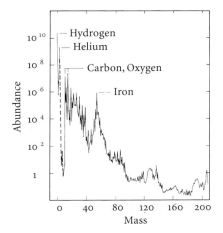

Fig. 1.7. Composition of the Sun. The lightest element, hydrogen, is the most abundant element in the Sun, and helium is the next most abundant one. The exponential decline of abundance with increasing mass can be explained if all the elements heavier than helium were synthesized in the interiors of stars that no longer shine, and then wafted or blasted out into interstellar space. Notice that this abundance scale is logarithmic and spans twelve orders of magnitude, or one million million.

Adjacent lines of the same atom exhibit a strange regularity – they systematically crowd together and become stronger at shorter wavelengths. These regular spacings can only be explained if electrons occupy orbits with very specific, quantized energies.

Thus, electrons revolve about the nucleus according to very specific rules, and are restricted to orbits with definite, quantized values of energy. If an electron wants to stay inside an atom, it has to obey the rules and can only occupy well-defined orbits with fixed energies.

Moreover, because only quantized orbits are allowed, spectral lines are only produced at specific wavelengths that characterize or identify the atom. An atom or molecule can absorb or emit a particular type of sunlight only if it resonates to that light's energy. As it turns out, the resonating wavelengths or energies of each atom are unique – they fingerprint an element, encode its internal structure and identify the ingredients of the Sun. (In addition, spectral lines yield information about the Sun's temperature, density, motion and magnetism.)

Each element, and only that element, produces a unique set of absorption or emission lines. The presence of these spectral signatures can therefore be used to specify the chemical ingredients of the Sun (Fig. 1.7). The lightest element, hydrogen, is the most abundant element in the Sun and most other stars (see Focus 1A Composition of the Stars.) Altogether, 92.1 percent of the atoms in the Sun are hydrogen atoms, 7.8 percent are helium atoms, and all the other heavier elements make up only 0.1 percent of the atoms in our home star.

The Sun is mainly composed of light elements, hydrogen and helium, which are terrestrially rare, whereas the Earth is primarily made out of heavy elements that are relatively uncommon in the Sun. Hydrogen is, for example, about one million times more abundant than iron in the Sun, but iron is one of the main constituents of the Earth which cannot even retain hydrogen gas in its atmosphere.

Helium, the second-most abundant element in the Sun, is so rare on Earth that it was first discovered on the Sun – by the French astronomer Jules Janssen and the British astronomer Joseph Lockyer as emission lines observed during the solar eclipse of 1868. Since it seemed to be only found on the Sun, Lockyer named it after the Greek Sun god, *Helios,* who daily travelled across the sky in a chariot of fire drawn by four swift horses. In 1895, while analyzing a gas given off by a uranium compound, the prominent British scientist Sir William Ramsey found the spectral signature of helium, thereby isolating it on the solid Earth 27 years after its discovery in the Sun. Today, helium is used on Earth in a variety of ways, including inflating party balloons and in its liquid state keeping sensitive electronic equipment cold.

1.7 CHILDREN OF THE STARS

We are made of the same atoms as the Sun. Our bodies, like the Sun, have more hydrogen atoms than any other, but we are composed of a somewhat larger proportion of heavier elements like carbon, nitrogen, and oxygen.

But do not discount the other stars. We are all true children of the stars, partially composed of materials that were forged within ancient stars before the Sun was born. All of the elements heavier than helium were generated long, long ago and far, far away in the nuclear crucibles of other stars.

These stars used up their internal fuel and spewed out their cosmic ashes with explosive force, ejecting the heavier elements into interstellar space. From this recycled material, the Sun, the Earth and we ourselves were formed. So, the calcium in your teeth, the sodium in your salt, and the iron that reddens your blood came from the interiors of other stars, long since exploded back into space in the death throes of these stars.

1.8 DESCRIBING THE RADIATION

The Sun continuously radiates energy that spreads throughout space. This radiation is called "electromagnetic" because it propagates by the interplay of oscillating electrical and magnetic fields in space. Electromagnetic waves all travel through empty space at the same constant speed – the velocity of light. (This velocity is usually denoted by the lower case letter c, and it has a value of roughly 300 000 kilometers per second – a more exact value is 299 793 kilometers per second.) No energy can be transported more swiftly than the speed of light.

Sunlight has no way of marking time, and it can persist forever. As long as a ray of sunlight passes through empty space and encounters no atoms or electrons it will survive unchanged. Radiation emitted from the Sun today might therefore travel for all time in vacuous space, bringing its message forward to the end of the Universe.

Some of the radiation streaming away from the Sun is nevertheless intercepted at Earth, where astronomers describe it in terms of its wavelength, frequency or energy. When light propagates from one place to another, it often seems to behave like waves or ripples on a pond (Fig. 1.8). The light waves have a characteristic wavelength, the separation between adjacent wave crests.

Different types of electromagnetic radiation differ in their wavelength, although they propagate at the same speed. The electromagnetic waves entering your eye and those picked up by your radio or TV antenna are similar except in respect to their wavelength. (Radio waves are too long to enter the eye, and not energetic enough to affect vision.)

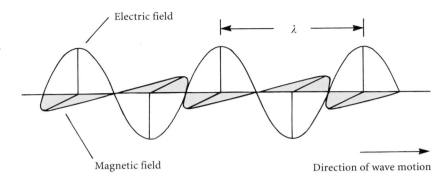

Fig. 1.8. Electromagnetic Waves. All forms of radiation consist of electric and magnetic fields that oscillate at right angles to each other and to the direction of travel. They move through empty space at the velocity of light. Like waves on water, the electromagnetic waves have peaks and troughs. The separation between consecutive peaks or troughs is called the wavelength of the radiation and is usually designated by the lower case Greek letter lamda or λ.

Just as a source of sound can vary in pitch, or wavelength, depending on its motion, the wavelength of electromagnetic radiation shifts when the body emitting or reflecting the radiation moves with respect to the observer. This is called the Doppler effect, after Christian Doppler who discovered it in 1842. If the motion is toward the observer, the Doppler effect shifts the radiation to shorter wavelengths, and when the motion is away the wavelength becomes longer. Such an effect is responsible for the change in the pitch of a passing ambulance siren. Because everything in the Universe moves, the Doppler effect is a very important tool for astronomers; it measures the velocity of that motion along the line of sight to the observer.

Sometimes radiation is described by its frequency instead of its wavelength (Fig. 1.9). Radio stations are, for example, denoted by their call letters and the frequency of their broadcasts, usually in units of a million cycles per second, or megahertz, for FM broadcasts.

The frequency of a wave is the number of crests passing a stationary observer each second; the frequency therefore tells us how fast the radiation oscillates, or moves up and down. The product of wavelength and frequency equals the velocity of light, so when the wavelength increases the frequency decreases and *vice versa*.

When light is absorbed or emitted by atoms, it behaves like packages of energy, called photons. The photons are created whenever a material

Fig. 1.9. Frequency Spectrum. The frequencies of electromagnetic waves (*top*) are compared with the spectral region (*middle*), and their common use on Earth (*bottom*). The top frequency scale is further divided into the low-frequency range that is not energetic enough to ionize an atom, and the high-frequency one where radiation has sufficient energy to strip one or more electrons from an atom, thereby forming an ion.

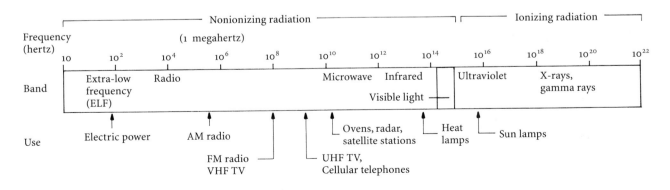

object emits electromagnetic radiation, and they are consumed when radiation is absorbed by matter. Moreover, each elemental atom can only absorb and radiate a very specific set of photon energies.

At the atomic level, the natural unit of energy is the electron volt, or eV. One electron volt is the energy an electron gains when it passes across the terminals of a 1-volt battery. A photon of visible light has an energy of about two electron volts, or 2 eV. Much higher energies are associated with nuclear processes; they are often specified in units of millions of electron volts, or MeV. A somewhat lower unit of energy is a thousand electron volts, called the kilo-electron volt, or keV; it is often used to describe X-ray radiation.

The interaction of each type of radiation with matter depends on the energy of its photons, and from the standpoint of the astrophysicist this is the most important property distinguishing one type of radiation from another. In fact, astronomers often describe energetic radiation, such as X-rays or gamma rays, in terms of its energy rather than its wavelength or frequency.

Photon energy is inversely proportional to the wavelength and directly proportional to the frequency. Radiation with a shorter wavelength or a higher frequency therefore corresponds to photons with higher energy. Radio photons have relatively long wavelengths and low frequencies, so they have small energy compared to the short-wavelength, high-frequency X-ray radiation. The small energies of radio photons cannot easily excite the atoms of our atmosphere, so these photons easily pass through it. In contrast, X-rays are totally absorbed when travelling just a short distance through the air. It also takes much less energy to broadcast a radio signal over short distances than to take an X-ray of a broken bone.

The energy of the radiation at a given wavelength can be related to the thermal energy, or the temperature, of the emitting gas. Hot stars tend to be bluer in color, for example, and colder stars are redder. This is because the most intense emission occurs at a radiation frequency and energy that increase with the star's surface temperature. In other words, the emitted power peaks out at a wavelength that varies inversely with the temperature, and this applies to all gaseous objects in the cosmos. Thus, the dark, cold spaces between the stars radiate most intensely at the longer, invisible radio wavelengths, while a hot, million-degree gas emits most of its energy at short X-ray wavelengths that are also invisible.

1.9 INVISIBLE FIRES

There is more to the Sun than meets the eye! In addition to visible light, there are invisible gamma ray, X-ray, ultraviolet, infrared and radio waves. The whole solar spectrum extends from short gamma rays, that are comparable to the size of an atom's nucleus, to long radio waves that are as broad as a mountain; and the Sun is so bright that it can be examined with precision in every spectral region. Observations at these invisible wavelengths have indeed broadened and sharpened our vision of the Sun.

However, our atmosphere effectively blocks most forms of invisible radiation including ultraviolet and X-ray radiation (Fig. 1.10). Radio waves are the only kind of invisible radiation that is not absorbed in the Earth's atmosphere, so radio astronomy provided the first new window on the Sun.

Astronomers use conventional radio telescopes to observe the Sun (Fig. 1.11); but radio telescopes do not really look at the Sun, they listen to it. Such telescopes usually have a metallic, dish-shaped, or parabolic, reflector that focuses the radio waves at a receiver. The long, straight antenna on your automobile or home radio similarly intercepts radio signals. Moreover, the Sun is the brightest, noisiest radio object in the sky, and because the atmosphere does not distort radio signals we can observe the radio Sun on a cloudy day, just as your home radio works even when it rains or snows outside.

To look at the Sun through windows other than the radio or visible ones, we must loft telescopes above the atmosphere. This was done first

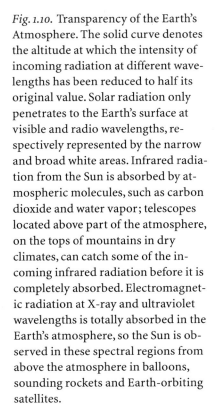

Fig. 1.10. Transparency of the Earth's Atmosphere. The solid curve denotes the altitude at which the intensity of incoming radiation at different wavelengths has been reduced to half its original value. Solar radiation only penetrates to the Earth's surface at visible and radio wavelengths, respectively represented by the narrow and broad white areas. Infrared radiation from the Sun is absorbed by atmospheric molecules, such as carbon dioxide and water vapor; telescopes located above part of the atmosphere, on the tops of mountains in dry climates, can catch some of the incoming infrared radiation before it is completely absorbed. Electromagnetic radiation at X-ray and ultraviolet wavelengths is totally absorbed in the Earth's atmosphere, so the Sun is observed in these spectral regions from above the atmosphere in balloons, sounding rockets and Earth-orbiting satellites.

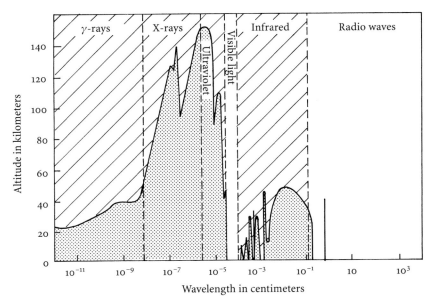

by using balloons and sounding rockets, and then by satellites that orbit the Earth above the atmosphere. Satellite-borne telescopes now view the Sun at ultraviolet and X-ray wavelengths, above the Earth's absorbing atmosphere in a world where night is brief.

Thus, astronomers now have new ways to extract previously-unobtainable information about the Sun. They are aided by new telescopes and sophisticated computers that gather in the increasing wealth of unsuspected information. Computerized telescopes now operate, from the ground and in orbit, in each of the invisible domains of the electromagnetic spectrum, creating images that provide new insight to the Sun.

Much of this book describes this invisible Sun, an unseen world of perpetual change and cosmic violence that lies outside the visible surface of the Sun. And as the title *Sun, Earth and Sky* suggests, our book also describes the Sun's interaction with planet Earth, mainly through invisible radiation and tiny, energetic particles that cannot be seen. It involves a global, space-age perspective that looks up at the Sun and down at the

Fig. 1.11. Rainbow above the VLA. The Very Large Array, abbreviated VLA, operates at radio wavelengths. In this photograph, it is strikingly portrayed against the colors of a double rainbow. The VLA is a collection of radio telescopes interconnected electronically to provide a total of 351 pairs of telescopes. The combined signals obtain two hundred thousand pieces of information every hour, so giant computers are also required to carry out radio investigations of the Sun. (Courtesy of Douglas Johnson, Batelle Observatory, Washington)

Earth, at both visible and invisible wavelengths, or directly samples the space outside Earth with orbiting satellites.

But before we begin our journey through these largely invisible realms, there is an equally fascinating world that lies hidden below everything we can see on the Sun; clever techniques are required to perceive this invisible interior of the Sun.

Impression of the Rising Sun. 1872.
This impression of sunrise by Claude
Monet gave the Impressionist Move-
ment its name. (Courtesy of the
Musée Marmottan, Paris)

CHAPTER 2

Energizing the Sun

2.1 AWESOME POWER, ENORMOUS TIMES

The Sun is relentlessly losing its energy, radiating it away at an enormous rate. In just one second, the Sun emits more energy than humans have used since the beginning of civilization. Its fire is too brilliant to be perpetually sustained; after all, nothing can stay hot forever, and all things wear out with time.

Why does the Sun stay hot? A normal fire of the Sun's intensity would soon burn out. That is, no ordinary fire can maintain the Sun's steady supply of heat for long periods of time. If the Sun was composed entirely of coal, with enough oxygen to sustain combustion, it could only generate as much heat and light as the Sun now does for a few thousand years. Even a Sun-sized lump of coal would then be burned away and totally consumed.

William Thomson, later Lord Kelvin, showed that the Sun could have illuminated the Earth at its present rate for a much longer time, about 100 million years, by slowly contracting. If the Sun was gradually shrinking, the infalling matter would collide and heat the solar gases to incandescence, just as the air inside a tire pump warms when it is compressed; in more scientific terms, the Sun's gravitational energy would be slowly converted into the kinetic energy of motion and heat the Sun up. (Thomson developed the scale of temperature that starts from absolute zero – the temperature at which atoms and molecules cease to move and have no kinetic energy; it is now known as the Kelvin temperature scale and is widely used by astronomers.)

In his article entitled "On the Age of the Sun's Heat", published in 1862, Thomson wrote:

> It seems, therefore, on the whole most probable that the Sun has not illuminated the Earth for 100 000 000 years, and almost certain that he has not done so for 500 000 000 years. As for the future, we may say, with equal certainty, that inhabitants of the Earth cannot continue to enjoy the light and heat essential to their life, for many million years longer, unless sources now unknown to us are prepared in the great storehouse of creation.[5]

During the ensuing decades, radioactivity was discovered, leading to the realization that the Earth's rocks are older than Thomson's value for the age of the Sun's heat. This paradox was resolved when a new source of heat was discovered – nuclear power.

Most of the matter on Earth is completely stable, but some atoms are unstable. Such radioactive atoms, like uranium, spontaneously change form when their nucleus hurls out energetic particles, radiates energy and relaxes to a less energetic state, forming a lighter, stable atom, like lead, in the process. This nuclear transformation can be used to determine the age of the rocks on the Earth's surface.

The radioactive dating method is something like determining how long a log has been burning by measuring the amount of ash and waiting a while to determine how rapidly the ash is being produced. Except you do not need to know the total amount of radioactive ash. The abundance ratio of stable decay atoms to their unstable parents, such as the relative amounts of lead and uranium, can be used with the known rate of radioactive decay to determine the time that has elapsed since the rocks were formed. This technique indicates that the oldest known rocks in the solar system (from the Moon and meteorites) were formed 4.6 billion years ago, when the Sun is thought to have been born.

Fossils of primitive creatures are found etched in rocks more than 3 billion years old, so the Sun was apparently warm enough to sustain life back then. Unusual powers must be at work to make the Sun shine so brightly for so long. Indeed, the only known process that can fuel the Sun's fire at the presently observed rates for billions of years involves nuclear fusion in the Sun's hot, dense core.

2.2 THE SUN'S CENTRAL PRESSURE COOKER

Under the extreme conditions within the Sun, atoms lose their identity! The atoms move rapidly here and there, colliding with each other at high speeds; the violent force of these collisions is enough to fragment the atoms into some of their constituent pieces. The interior of the Sun therefore consists mainly of the nuclei of hydrogen atoms, called protons, and free electrons that have been torn off the atoms by innumerable collisions and set free to move throughout the Sun's core.

Negatively charged electrons neutralize the positively charged protons, so the mixture of electrons and protons, called a plasma, has no net charge. The entire Sun is nothing but a giant, hot ball of plasma. (A plasma has been called the fourth state of matter to distinguish it from the gaseous, liquid, and solid ones.) A candle flame is a plasma.

With their electrons gone, the hydrogen nuclei, or protons, can be packed together much more tightly than normal atoms. This is because nuclei are 40 000 times smaller than the atoms they normally occupy. The bare nuclei can be squeezed together within the empty space of former atoms.

To understand the Sun's interior, imagine a hundred mattresses stacked into a pile. The mattresses at the bottom must support those above, so they will be squeezed thin. Those at the top have little weight to carry, and they retain their original thickness. The nuclei at the center of the Sun are similarly squeezed into a smaller volume by the overlying material, so they become hotter and more densely concentrated.

Deep down inside, within the dense, central core, the Sun's temperature has risen to more than 15 million degrees Kelvin and the gas density is greater than 10 times that of solid lead; see Table 2.1. (Such extreme central conditions were recognized as long ago as 1870 when Jonathan Homer Lane, an American physicist at the U.S. Patent Office, assumed that gas pressure supports the weight of the Sun.) As the result of such crowding, the nuclei in the Sun's center collide more frequently with higher speeds than elsewhere in the Sun, and push more vigorously outward. This pushing is called gas pressure, and it is the force that keeps the Sun from collapsing. (Fast moving particles also emit electromagnetic radiation, resulting in an additional outward force called radiation pressure; it is important in supporting giant stars but not the Sun.)

At the center of the Sun, the gas pressure needed to resist the weight of the overlying gas is 300 billion times the pressure of our atmosphere at sea level. The enormous central pressure is provided by the high-speed motions and collisions of particles with temperatures of 15.6 million degrees!

Fig. 2.1. Internal Compression. The variation of pressure, luminosity, temperature and density with fractional radial distance from the Sun's center (*left*) to its visible surface (*right*). At the Sun's center, the temperature is 15.6 million degrees Kelvin and the density is 151 grams per cubic centimeter; the central pressure is 233 billion times that of the Earth's atmosphere at sea level (one bar). Thermonuclear energy production takes place in a core region extending to about one-quarter (0.25) of the solar radius; the core contains almost half of the Sun's mass. The convective zone begins at 0.71 of the solar radius where the temperature has dropped to about 2 million degrees Kelvin and the density has fallen to about 0.2 grams per cubic centimeter; the convective zone comprises about 2 per cent of the Sun's mass. At the photosphere, the temperature is 5780 degrees Kelvin, and the pressure and density have dropped off the scales of the graph. [Adapted from the standard solar model computed by John Bahcall and Marc H. Pinsonneault, *Reviews of Modern Physics*, *64*, 885–926 (1992)]

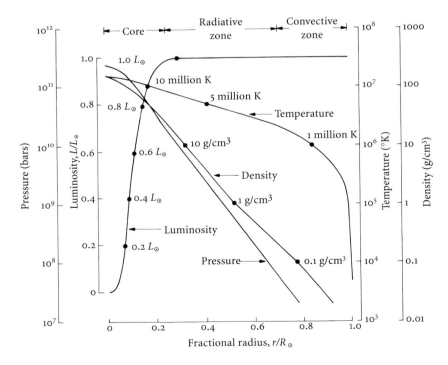

Table 2.1. The Sun's vital statistics[*]

Radius, $R_\odot$	6.9598×10^{10} cm (109 Earth radii)
Mass, $M_\odot$	1.989×10^{33} grams (3×10^5 Earth masses)
Luminosity, $L_\odot$	3.854×10^{33} erg s^{-1}
Age	4.55×10^9 years
Volume	1.412×10^{33} cm^3 (1.3 million Earths)
Mean density	1.409 g cm^{-3}
Helium abundance, Y	0.28 ± 0.01 by mass
Central temperature	1.557×10^7 degrees Kelvin
Central density	151.3 g cm^{-3}
Central pressure	2.334×10^{17} dyn cm^{-2} (2.334×10^{11} bars)
Depth of convection zone	$0.287\,R_\odot$ (radius $0.713 \pm 0.003\,R_\odot$)
Mass of convection zone	$0.022 \pm 0.002\,M_\odot$
Temperature at base of convection zone	2.12 to 2.33×10^6 degrees Kelvin
Temperature of photosphere	5780 degrees Kelvin
Pressure of photosphere	10^{-4} bars
Temperature of corona	2 to 3×10^6 degrees Kelvin
Mean distance, A.U.	1.4959787×10^{13} cm
Angular diameter at mean distance	32 arc-minutes (1920 arc-seconds)
Angular scale	1 arc-second = 7.253×10^7 cm

[*] Adapted from K. R. Lang: Astrophysical Data, Planets and Stars. Springer, Berlin Heidelberg 1991; J. N. Bahcall and M. H. Pinsonneault, Reviews of Modern Physics *64*, 885–926 (1992); J. Christensen-Dalsgaard, et al., Astrophysical Journal *378*, 413–437 (1991); D. B. Guenther et al., Astrophysical Journal *387*, 372–393 (1993); I.-J. Sackmann et al., Astrophysical Journal *418*, 457–468 (1993). The radius of the photosphere is at an optical depth of $\tau = 2/3$, the mass uncertainty is $\pm\,0.02$ %, the luminosity uncertainty is 1.5 % with a range of 3.846 to 3.857 from measurements of the solar constant of 1.368 to 1.378, the age of the oldest meteorite is estimated to be 4.55 billion years, but perhaps as old as 4.6 to 4.7 billion years, the Sun's age from evolutionary calculations is 4.52 ± 0.4 billion years, oscillation data indicate that the helium abundance in the convection zone is $Y = 0.23$ to 0.26, the value of Y given here is from the standard solar model, and the central temperature, density and pressure are from the standard solar model without helium diffusion.

At greater distances from the center, there is less overlying material to support and the compression is less, so the plasma gets thinner and cooler (see Fig. 2.1). Halfway from the center of the Sun to the surface, the density is the same as that of water, only 1 gram per cubic centimeter; about nine tenths of the distance from the center to the Sun's apparent edge, we find material as tenuous as the transparent air that we breathe on Earth. At the visible surface of the Sun, the rarefied solar gas is about ten thousand times less dense than our air, the pressure is less than that beneath the foot of a spider, and the temperature has fallen to 5780 degrees Kelvin.

2.3 NUCLEAR FUSION, ANTI-MATTER AND HYDROGEN BURNING

The extraordinary conditions within the center of the Sun provided one clue to the mysterious process that keeps the Sun hot and makes it shine. Other important evidence was accumulated at the Cavendish Laboratory in Cambridge, England; first when Ernest Rutherford showed that radioactivity is associated with nuclear transformation and then when Francis Aston demonstrated that the mass of the helium atom is slightly less than the sum of the masses of the four hydrogen atoms which enter into it. (Rutherford and Aston respectively won the Nobel Prize for their work in 1908 and 1922.)

At about the same time, the British astronomer, Arthur Stanley Eddington was trying to understand the inner workings of the Sun and other stars. Eddington, an avid reader of mystery novels, once likened the process to analyzing the clues in a crime. He knew that certain elements can be transformed into other ones in the terrestrial laboratory, and reasoned that stars are the crucibles in which the elements are made. He further realized that such stellar alchemy would release energy, arguing that hydrogen is transformed into helium inside stars, with the resultant mass difference released as energy to power the Sun.

Eddington could therefore lay the foundation for solving the Sun's energy crisis, concluding in 1920 that:

> What is possible in the Cavendish Laboratory may not be too difficult in the Sun .... The reservoir [of a star's energy] can scarcely be other than the subatomic energy .... There is sufficient [subatomic energy] in the Sun to maintain its output of heat for 15 billion years.[6]

In the same article, he continued with the prescient statement that:

> If indeed, the subatomic energy in the stars is being freely used to maintain their great furnaces, it seems to bring a little nearer to fulfillment our dream of controlling this latent power for the well-being of the human race – or for its suicide.[7]

Great ideas have a curious way of surfacing in different places at about the same time. In an essay entitled "Atoms and Light", the French physicist Jean Perrin argued that "radioactive" transformation of the elements could maintain the Sun's luminous output at its present rate for several billion years, or perhaps several dozen billion years, and that the mass lost, m, during the transformation of four hydrogen nuclei into one helium nucleus would supply energy, E, according to Einstein's equation $E = mc^2$, where c is the velocity of light.[8]

It was probably Eddington who convinced most astronomers that subatomic (that is, nuclear) energy must fuel the stars. During the ensu-

ing decade it was realized that the lightest known element, hydrogen, is
the most abundant element in the Sun, so hydrogen nuclei, or protons,
must play the dominant role in nuclear reactions within our daytime
star. Protons must somehow be fused together, forming helium nuclei,
but the details were lacking. After all, Rutherford had only shown that
hydrogen has a proton nucleus in 1919, the year before Eddington's his-
toric article, and the subatomic world was largely unknown.

Physicists were nevertheless convinced that protons could not react
with each other inside the Sun. The main problem in accepting their fusion was that protons resist coming together because of the repulsion of
their positive electrical charges, and this repulsive electrical barrier cannot be overcome in a head-on collision of protons in the Sun, even at its
enormous central temperature. In other words, the temperature at the
center of the Sun was thought to be far too low to permit protons to fuse
together and for nuclear reactions to occur.

But Eddington was certain that subatomic energy fueled the stars,
and in the mid-1920s retorted defiantly that:

> The helium which we handle must have been put together at some
> time and some place. We do not argue with the critic who urges that
> the stars are not hot enough for this process; we tell him to go and
> find *a hotter place*.[9]

As it turned out, Eddington was right and the physicists were wrong.

(It was later realized that most of the helium we now see was synthesized in the high-temperature, big-bang explosion that gave rise to the
expanding Universe, but that lesser amounts of helium are also manufactured inside stars.)

The paradox was resolved with the discovery of quantum mechanics,
which combines the wave and particle aspects of matter; neither aspect
individually describes events at the atomic or subatomic level. In this uncertain quantum world, the magic of probability comes into play and a
particle's sphere of influence is larger than was previously thought.

A proton acts like a spread-out wave, with no precisely defined position; and the proton's energy fluctuates about its average, thermal value.
This means that it has a very small but finite chance of occasionally moving close enough to another proton to overcome the barrier of repulsion
and tunnel through it. (A similar tunneling process occurs the other way
around when radioactive nuclei, like radium or uranium, rid themselves
of excess protons by throwing out high-speed particles; these particles
lack the energy to overcome the nuclear barrier, but some of them tunnel
through to the outside world.)

In this surreal world of subatomic probability, one could relentlessly
throw a ball against a wall, watching it bounce back countless times, until
eventually the ball would go through (or under) the wall. As Ahab said in
Moby Dick,

How can the prisoner reach outside except by thrusting through the wall?[10]

Thus, protons sometimes get close enough to fuse together, even though their average energy is well below that required to overcome their electrical repulsion. But this bizarre tunneling reaction doesn't occur all the time. For fusion to occur, the collision must still be almost exactly head-on, and between exceptionally fast protons. Nuclear reactions therefore proceed very slowly inside the Sun, and it is a good thing. If the temperature was high enough to permit frequent fusion, the Sun would blow up! After all, similar nuclear processes produce the explosive energy in hydrogen bombs.

Unlike a bomb, the temperature-sensitive reactions inside the Sun act like a thermostat, releasing energy in a steady, controlled fashion at exactly the rate needed to keep the Sun in equilibrium. If a star shrinks a little and gets hotter inside, more nuclear energy is generated, making the star expand and restoring it to the original temperature. If the Sun expanded slightly, and became cooler inside, subatomic energy would be released at a slower rate, making the Sun shrink again and restoring equilibrium.

So, we now know that the Sun shines by nuclear fusion, whereby hydrogen nuclei, or protons, fuse together into helium nuclei, also known as alpha particles. The detailed sequence of nuclear reactions is known as the proton-proton chain (see Focus 2A), since it begins by the fusion of two protons. It was delineated by Hans A. Bethe and his colleagues in the late 1930s.

FOCUS 2A
Proton-Proton Chain

The Sun shines by a sequence of nuclear reactions, called the proton-proton chain, in which four protons are fused together to form a helium nucleus that contains two protons and two neutrons. Each nuclear transformation releases 25 MeV, or 0.000 04 erg, of energy. This results from the fact that the mass of the resulting helium nucleus is slightly less (a mere 0.007 or 0.7 percent) than the mass of the four protons that formed it, and the missing mass appears as energy.

The energy content of the lost mass is given by $E = mc^2$. Because the velocity of light, c, is a very large number, the annihilation of relatively small amounts of mass, m, produces large quantities of energy, E. Moreover, that energy is multiplied by the huge number of reactions that occur inside the Sun every second. Roughly 100 trillion trillion trillion, or ten to the thirty-eight, helium nuclei are created every second, resulting in a total mass loss of 5 million tons per second, which is enough to keep the Sun shining with its present brilliant luminous output of 4 million, billion, billion, billion, or 4 times ten to the thirty-three, ergs per second.

In the first step of the proton-proton chain, two protons, p, are united to form a deuteron, D^2, the nucleus of a heavy form of hydrogen known as deuterium. Since a deuteron consists of one proton and one neutron, one of the protons entering the reaction must be transformed into a neutron, emitting a positron, e^+, to carry away the proton's charge, together with a low-energy neutrino, ν_e, to balance the energy in the reaction. This initiating proton-proton, or p–p, reaction is written:

$$p + p \rightarrow D^2 + e^+ + \nu_e \ . \tag{2.1}$$

Each proton inside the Sun is involved in a collision with other protons millions of times in every second, but only exceptionally hot and lucky ones are able to tunnel through their electrical repulsion and fuse together. Just one collision in every ten trillion trillion initiates the proton-proton chain.

Part of the energy liberated during this reaction is converted into radiation when an electron, e^-, collides with a positron, e^+. Then matter and anti-matter annihilate each other to produce pure energy at short gamma ray wavelengths, γ. This radiation-producing interaction can be written as:

$$e^- + e^+ \rightarrow 2\gamma \ . \tag{2.2}$$

The next step follows with little delay. In less than a second the deuteron collides with another proton to form a nucleus of light helium, He^3, and releases yet another gamma ray.

$$D^2 + p \rightarrow He^3 + \gamma \ . \tag{2.3}$$

(This reaction occurs so easily that deuterium cannot be synthesized inside stars; it is quickly consumed to make heavier elements.)

In the final part of the proton-proton chain, two such light helium nuclei meet and fuse together to form a nucleus of normal heavy helium, He^4, returning two protons to the solar gas; this step takes about a million years on average.

$$He^3 + He^3 \rightarrow He^4 + 2p \ . \tag{2.4}$$

(This normal helium nucleus contains two protons and two neutrons.)

A total of six protons are required to produce the two He^3 nuclei that go into this last reaction. Since two protons and a helium nucleus are produced, the net result of the proton-proton chain is:

$$4p \rightarrow He^4 + \text{gamma ray radiation} + 2 \text{ neutrinos} \ . \tag{2.5}$$

Although Bethe was familiar with recent developments in nuclear physics, he was unacquainted with the astrophysical problems of stellar energy sources until April 1938 when George Gamow organized a conference in Washington D. C., to bring astronomers and physicists together to discuss this problem. Bethe was so stimulated by the meeting that he acquired the necessary astrophysical knowledge and specified the various nuclear reactions that make both the Sun and other hydrogen-burning stars shine. (Bethe received the Nobel Prize in 1967 for his discoveries concerning the energy production in stars.)

In the burning of hydrogen, four protons are united, but two of them have to be changed into neutrons. This is because the helium nucleus consists of two protons and two neutrons. Something must be carrying away the charge of the proton, leaving behind a neutron with little change of mass; that mysterious agent is the anti-particle of the electron.

Anti-matter was predicted in 1930 by Paul A. M. Dirac, then at the University of Cambridge. For Dirac, mathematical beauty was the most important aspect of any physical law describing nature. He noticed that the equations that describe the electron have two solutions. Only one of them was needed to characterize the electron; the other solution specified a sort of mirror-image of the electron – an anti-particle, now called the positron.

Dirac's trust in the beauty and symmetry of his equations led him to predict:

> a new kind of particle, unknown to experimental physics, having the same mass and opposite charge to an electron.[11]

The then-unknown positron was discovered in 1932 by Carl D. Anderson when studying high-energy particles from space, called cosmic rays; they create positrons and many other subatomic particles when colliding with nuclei in our outer atmosphere. (Dirac received the Nobel Prize in 1933 for his prediction; Anderson received the honor in 1936 for the discovery of the positron, in the same year as Victor Hess for his discovery of cosmic rays.)

Once created, anti-matter does not stay around for very long. And it is a good thing, for we occupy a material world, and any anti-matter will self-destruct when it encounters ordinary matter. The singer Madonna expressed it with a different connotation:

> We are living in a material world
> And I am a material girl.[12]

When an electron and positron meet, they annihilate each other and disappear in a puff of energetic radiation. This is how some of the Sun's nuclear energy is converted into radiation. (The characteristic radiation produced during the annihilation of electrons and positrons, the 0.511 MeV line, has been detected during powerful eruptions on the Sun called solar flares.)

As Eddington suggested, the actual source of the Sun's energy is the annihilation of mass, through Einstein's celebrated equation $E = mc^2$. Mass, m, and energy, E, are not distinct and permanent constituents of the Universe; they are fully interchangeable and the abolition of mass releases energy.

Every time four protons are fused together to make a helium nucleus, the mass loss results in the release of about 25 million electron volts, or 25 MeV, of thermal energy. By itself, that is not very much energy, but when multiplied by the huge number of reactions that occur every second, it's enough to keep the Sun shining. Thus, the Sun is losing weight in order to shine brightly – at the rate of 5 million tons each second.

Although the Sun is consuming itself at a prodigal rate, the loss of material is insignificant in comparison with its total mass. (The mass of the Sun is two thousand trillion trillion tons, or about a third of a million times the mass of the Earth – see Table 2.1.) Over the past 4.6 billion years, the Sun has consumed only a few hundredths of one percent of its original mass. The reason is essentially because very little mass (a mere 0.007 or 0.7 percent) is annihilated in forming a helium nucleus.

And, on the average, very little heat is actually released by each gram of matter in the Sun – much less than the rate that heat is released in your body. This is because solar heat escapes from a relatively small surface area compared to the enormous volume of the Sun. A massive elephant similarly retains internal heat so efficiently that its large ears have to help radiate the heat away, whereas a tiny mouse easily loses heat and has to eat all the time just to stay warm inside. (A mouse heart beats during a mouse lifetime about as many times as an elephant's does in its lifetime.)

A more significant concern is the depletion of the Sun's hydrogen fuel within its nuclear furnace. About 600 million tons of hydrogen are transformed into helium every second to make the Sun shine as brightly as it does. But only a small fraction of that mass, about 5 million tons, is released as energy; most of the hydrogen is burned into helium ash. About 37 percent of the hydrogen originally in the Sun's core has been converted into helium. Since thermonuclear reactions are limited to the hot, dense core, it will eventually run out of hydrogen – in about 5 billion years. The Sun will then expand to engulf the Earth.

Fig. 2.2. Anatomy of the Sun. The Sun ▷ is an enormous ball of gas. Its interior consists of two nested shells surrounding an energy-generating core. The central gaseous material is compressed to such high densities and temperatures that thermonuclear reactions can take place there. Energy is transported by radiation within the overlying radiative zone. In the outer convective zone, energy is carried by the turbulent motion of a hot gas. The radiative transport of energy briefly takes over again in the thin visible surface of the Sun, or photosphere. The Sun maintains an enormous temperature difference between the 15-million-degree core and its 5780 degree visible disk. A million-degree corona envelops the photosphere; magnetism molds this outer atmosphere, confining the hot material in coronal loops and cooler material in prominences. Coronal holes offer escape routes for the high-speed solar wind. Courtesy of NASA)

1 Chromosphere
2 Spicules
3 Radiative zone
4 Convective zone
5 Core
6 Corona
7 Photosphere
8 High speed solar wind
9 Coronal hole
10 Coronal high temperature 2 million degrees
11 Sunspots
12 Loop prominence

2.4 DILUTING THE RADIATION

All the nuclear energy is released deep down inside the Sun's high-temperature core, and no energy is created in the cooler regions outside it. The energy-generating core extends to about one quarter of the distance from the center of the Sun to the visible surface, accounting for only 1.6 percent of the Sun's volume. But about half the Sun's mass is packed into its dense core.

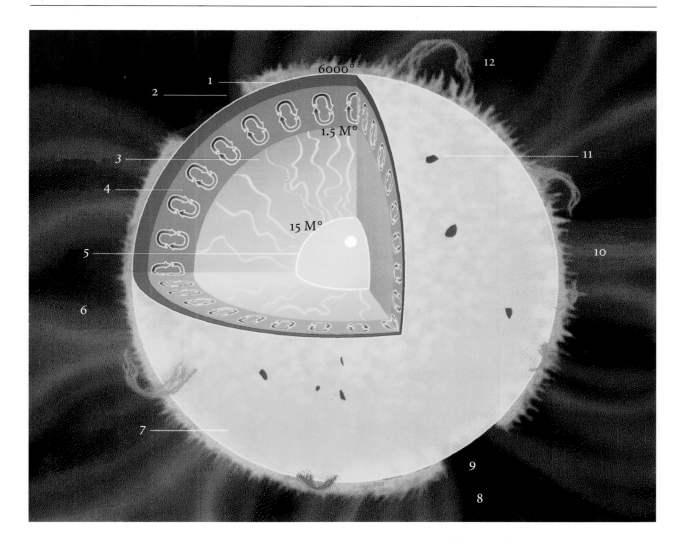

Because we cannot see inside the Sun, astronomers combine basic theoretical equations, such as those for equilibrium and energy generation or transport, with observed boundary conditions, such as the Sun's mass and luminous output, to create models of the Sun's internal structure. These models consist of two nested spherical shells that surround the hot, dense core, like Russian dolls (Fig. 2.2). In the innermost shell, called the radiative zone, energy is transported by radiation; it reaches out from the core to about 71 percent of the distance from the center of the Sun to the visible surface. The radiative zone is encompassed by a higher layer known as the convective zone, where a turbulent motion, called convection, transports energy.

Even though light is the fastest thing around, radiation does not move quickly from the center of the Sun to its visible surface. The energy made inside the Sun's pressurized core slowly trickles out to finally escape as the light that we see. The high-energy radiation collides with the plasma in the radiative zone, ricocheting randomly about and becoming absorbed, re-radiated and deflected, over and over again countless times.

Each time the radiation is re-emitted, it travels on a divergent course in a different direction, including back the way it came. However, the decrease in temperature with increasing distance from the Sun's center ensures that more radiation works its way outward than inward, just as heat normally flows from a hotter region to a colder one. In its random odyssey, the radiation therefore meanders toward the Sun's surface, following the path of least resistance like water flowing downhill, and heading for regions of lower density and temperature.

The radiation moves outward on a haphazard, zig-zag path, like a drunk staggering through a crowd of people, steadily losing energy and becoming exhausted. The gamma rays emitted during nuclear fusion, for example, cannot move more than one tenth of a centimeter before colliding with one of the many electrons in the core, producing shorter X-rays with slightly less energy. (Both the gamma rays and X-rays have relatively high energies and short wavelengths, and are invisible.) At each encounter with the plasma in the radiative zone, the solar radiation downshifts to lower energy and the wavelength becomes longer. By the time the radiation reaches the visible photosphere, the X-rays have changed through innumerable collisions to ultraviolet radiation, and finally to visible sunlight.

Because of this continued ricocheting in the radiative zone, it takes about 170 thousand years, on the average, for radiation to work its way out from the Sun's core to the bottom of the convective zone. (Energy is transported relatively quickly by convection.) Thus, the sunlight that we now see was generated inside the Sun before Neanderthalers wandered the globe. In contrast, sunlight moves freely through interplanetary space, taking only eight minutes to travel from the Sun to the Earth.

In the convection zone, stellar material is too cool and opaque to let the radiation pass. (The temperature has fallen to about 2 million degrees Kelvin at the bottom of the convection zone.) The cool ions absorb great quantities of radiation without re-emitting it. So, the radiative energy streaming out from the interior is blocked, like cars caught in a horrendous traffic jam. The material seethes with trapped energy, and the Sun must find another way to release the pent-up energy.

Huge convection currents, formed by hot material rising and cool material sinking, take over and transport heat outward. Travelling in essentially a straight line, large bubbles of ionized gas move outward through the convection zone at a modest speed, reaching the visible surface in roughly 10 days.

The convective zone resembles a boiling pot of water, with hot rising bubbles and cooler sinking material. The hot bubbles quickly rise to higher levels, cooling and expanding like hot air rising in the Earth's at-

Fig. 2.3. Bénard Convection. When a gas or liquid is heated uniformly from below, convection takes place in vertical cells, known as Bénard cells. Warmer material rises at the centers of the cells and cooler material falls around their boundaries. This figure shows the hexagonal convection pattern in a layer of silicone oil that is heated uniformly below and exposed to ambient air above; light reflected from aluminum flakes shows fluid rising at the center of each cell and descending at the edges. The regular motion, long duration, and polygonal shapes of these cells are somewhat distorted in the Sun's turbulent convection zone. (Courtesy of Manuel G. Velarde and M. Yuste, Universidad Nacional de Educación a Distancia, Madrid, Spain)

Fig. 2.4. Double, Double Toil and Trouble. When optical telescopes zoom in and take a detailed look at the visible Sun, they resolve a strongly-textured granular pattern. Hot granules, each about 1500 kilometers across, rise at speeds of 500 meters per second (1000 miles per hour), like supersonic bubbles in an immense, boiling cauldron. The rising granules burst apart, liberating their energy, and cool material then sinks downwards along the dark, intergranular lanes. Bright concentrations of magnetism sometimes illuminate the dark lanes. This photograph was taken at 4308 Angstroms (an Angstrom is one hundred millionths of a centimeter) with a 10-Angstrom interference filter, it has an exceptional angular resolution of 0.2 arc-seconds, or 150 kilometers at the Sun. (Courtesy of Richard Muller and Thierry Roudier, Observatoire du Pic-du-Midi et de Toulouse)

mosphere. When it becomes cooler than its surroundings, the gas sinks to become reheated and rise again. Thus, roiling currents of hot and cool gas create a churning motion that carries heat from the bottom to the top (Fig. 2.3).

The turbulent, rumbling convection boils endlessly away in a seething honeycomb of rising and falling gas. The largest currents of gas and heat apparently spawn a multitude of smaller ones, visible at the surface in white light photographs as a pattern of granulation (Fig. 2.4). (White

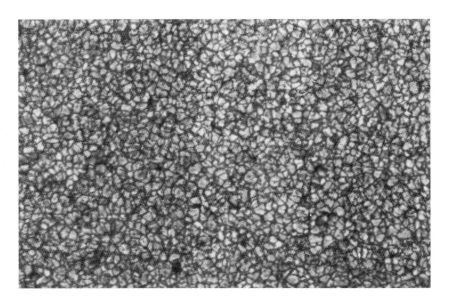

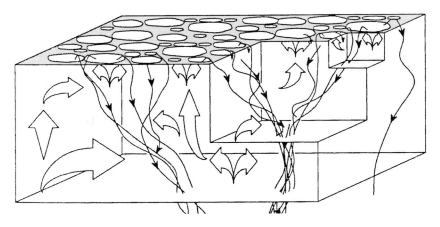

Fig. 2.5. Hierarchical Flows. Surface granular downflows merge into and drive larger scale ones at lower depths, like streams feeding a river. Thus, successively larger-scale convective flows are allowed at increasing depth below the Sun's visible surface. The boxes cut out illustrate how the same process occurs on (in this illustration) three different scales. [Adapted from H. C. Spruit, A. Nordlund and A. M. Title, Annual Review of Astronomy and Astrophysics *28,* 263–301 (1990), also see R. F. Stein and A. Nordlund, Astrophysical Journal (Letters) *342,* L95–L98 (1989)]

light is all the visible colors of solar radiation taken together.) The hot, bright granules measure about 1000 kilometers across and persist for only about ten minutes before dissolving to be replaced by new ones. They change minute by minute, indicating vigorous small-scale turbulence in the gases immediately below. Larger cells of horizontally moving material, called the supergranulation, are about 30 000 kilometers in diameter, suggesting circulation thought to be tens of thousands of kilometers deep. Theory suggests that convection in the deep interior is organized into even larger units, perhaps several hundred thousand kilometers in size (Fig. 2.5), but this has not been observed.

Radiation again takes over energy transport in the thin, outermost shell that we see; it is called the photosphere, which simply means the sphere from which the visible light comes (from the Greek *photos* for light). The photosphere is a layer only a few hundred kilometers thick, occupying less than a thousandth of the solar radius; it is a thin, tenuous opaque region.

The photosphere marks the visible surface of the Sun, but it is not really a surface. Indeed, the Sun has no definite surface that divides the inside from the outside. The entire Sun is just a gigantic, incandescent ball of gas, or plasma, that is compressed in the center and becomes more tenuous further out, extending to the Earth and beyond. For practical purposes, the photosphere is often taken to be the Sun's "surface", for most of the phenomena inside it are not sensitive to what is happening on the outside; but the opposite is not true.

In the photosphere, certain unusual ions control the radiation of visible sunlight into space. The photosphere is so cool that most of the free electrons have joined the protons to make hydrogen atoms. Yet, in a rare collision, a hydrogen atom in the photosphere can briefly attach a free electron to itself and temporarily become an ion with negative electrical charge. These negative hydrogen ions absorb radiative energy from the interior and re-emit visible sunlight.

Negative hydrogen ions absorb so much light that they dominate the transfer of radiation in the photosphere, even though their concentration is only a millionth that of the normal hydrogen atoms. In fact, the rarefied photosphere would be as transparent as the Earth's atmosphere if it were solely made of hydrogen atoms and did not contain any negative hydrogen ions.

Thus, the thin photosphere caps the convective zone and completes our current model of the solar interior. It is a simplified model that provides a good description of the Sun's global properties. Such a model might be expected to break down if we examined it in detail. Astronomers are now testing it with observations of oscillations generated deep within the Sun and neutrinos emitted from the Sun's energy-generating core.

Simultaneous Contrasts: Sun and Moon. 1912–13. In this painting, Robert Delaunay has portrayed the Sun with no reference to the earthly world. The Sun is seen as a source of pure light and color that can be appreciated in its own right without reference to the sunlight that illuminates terrestrial objects. (Courtesy of the Museum of Modern Art, New York. Mrs. Simon Guggenheim Fund, Oil on canvas, 53″ diameter)

CHAPTER 3

Ghostlike Neutrinos

3.1 THE ELUSIVE NEUTRINO

Neutrinos, or little neutral ones, are very close to being nothing at all. They are tiny, invisible packets of energy with no electric charge and little or no mass, travelling at or very near the velocity of light. These subatomic particles are so insubstantial, and interact so weakly with matter, that they streak through nearly everything in their path, like ghosts that move right through walls.

Unlike light or other forms of radiation, the neutrinos can move almost unimpeded through any amount of matter, even the entire Universe. Each second, trillions upon trillions of neutrinos that were produced inside the Sun pass right through the Earth without even noticing it is there. Billions of ghostly neutrinos from the Sun are passing right through you every second, whether you are indoors or outdoors, or whether it is day or night, and without your body noticing them, or them noticing your body.

The neutrinos are the true ghost riders of the Universe. As John Updike put it:

> Neutrinos, they are very small.
>   They have no charge and have no mass
> And do not interact at all.
> The Earth is just a silly ball
>   To them, through which they simply pass,
> Like dust maids down a drafty hall.[13]

How do we know that such elusive, insubstantial particles even exist? They are required by a fundamental principle, the conservation of energy. According to this rule, the total energy of a system must remain unchanged, unless acted upon by an outside force; we know of no process that disobeys this principle.

Nevertheless, in a process called beta-decay, the nucleus of a radioactive atom emits an energetic electron whose energy is less than that lost by the nucleus. Careful measurements failed to turn up the missing energy, which seemed to have vanished into thin air, and this suggested that energy may not be conserved during beta-decay. However, it turned out that a mysterious, invisible particle was spiriting away the missing energy.

Fig. 3.1. Wolfgang Pauli. This physicist predicted the existence of the neutrino to solve an energy crisis in a type of radioactivity called beta decay. Pauli thought that the invisible neutrino would never be seen, but it was subsequently discovered as a byproduct of nuclear reactions on Earth and in the Sun. (Courtesy of the American Institute of Physics Niels Bohr Library, Goudsmit Collection)

The neutrino's existence was therefore postulated more than half a century ago when Wolfgang Pauli, a brilliant Austrian physicist (see Fig. 3.1), proposed a "desperate way out" of the energy crisis. He speculated that a second, electrically neutral particle, produced at the same time as the electron, carried off the remaining energy. The sum of the energies of both particles remains constant, so the energy books are balanced during beta-decay, and the principle of conservation of energy is saved. As Pauli expressed it in 1933:

> The conservation laws remain valid, the expulsion of beta particles [electrons] being accompanied by a very penetrating radiation of neutral particles, which has not been observed so far.[14]

When first discovered, the electrons emitted during beta-decay were called beta-rays to distinguish them from alpha-rays (helium nuclei) and gamma-rays (high-energy radiation) that are also emitted during radioactive decay processes; from their measured charge and mass, we now know that beta-rays are not rays at all but instead ordinary electrons moving at nearly the velocity of light.

Pauli thought he had done "a terrible thing", for his desperate remedy postulated an invisible particle that could not be detected. Dubbed the neutrino, or "little neutral one" by the Italian physicist, Enrico Fermi, the new particle could not be observed with the technology of the day, which is not surprising considering that the neutrino is electrically neutral, has little or no mass, and moves at or nearly at the velocity of light. (Even in Pauli and Fermi's time, the observed high-energy shape of the emitted electron's energy spectrum indicated that the mass of the neutrino is either zero or very small with respect to the mass of the electron.)

As beautifully described by Fermi, the decay process occurs when the neutron in a radioactive nucleus transforms into a proton with the simultaneous emission of an energetic electron and a high-speed neutrino. When left alone outside a nucleus, a neutron will, in fact, self-destruct in about 10 minutes into a proton, plus an electron to balance the charge and a neutrino to help spirit away the energy.

As far as anyone could tell, an atomic nucleus consists only of more massive neutrons and protons, so the electron and neutrino seemed to come out of nowhere. They do not reside within the nucleus and are born at the time of nuclear transformation. No one knew exactly how the neutrinos were created.

How do you observe something that spontaneously appears out of nowhere and interacts only rarely with other matter? Calculations suggested that the probability of a neutrino interacting with matter, so one could see it, is so vanishingly small that one could never detect it. To see one neutrino, you would have to produce enormous numbers of them at about the same time, and build a very massive detector to increase the chances of catching it. Although almost all of the neutrinos would still pass through any amount of matter unhindered and undetected, a rare collision with other subatomic particles might leave a trace.

Nuclear reactors, first developed in the 1940s, produce large numbers of neutrinos. Such reactors run by a controlled chain reaction in which neutrons bombard uranium nuclei, causing them to split apart and create more neutrons to continue the chain reaction, and creating large amounts of energy with an enormous flux of neutrinos in the process. (The same thing occurs in an atomic bomb, except that the chain reaction runs out of control with an explosive release of energy.) If you place a very massive detector near a large nuclear reactor, and appropriately shield the detector from extraneous signals, you might just barely observe the tell-tale sign of the hypothetical neutrino.

The existence of the neutrino was finally proven with Project Poltergeist, an experiment designed by Frederick Reines and Clyde Cowan of the Los Alamos National Laboratory in New Mexico. They placed a 10-ton tank of water next to a powerful nuclear reactor engaged in making plutonium for use in nuclear weapons. After shielding the neutrino trap underground and running it for about 100 days over the course of a year, Reines and Cowan detected a few, tiny, synchronized flashes of gamma radiation that signaled the absorption of a few neutrinos by protons in the water. Then in June, 1956, they telegraphed Pauli with the news:

We are happy to inform you that we have definitely detected
neutrinos from fission fragments by observing inverse beta-decay
of protons.[15]

(The inverse beta-decay mentioned in the telegram occurs when a nuclear proton absorbs a neutrino and turns into a neutron, at the same time emitting a positron that immediately annihilates with an electron and produces the tell-tale burst of radiation that was detected.)

The ghostly neutrino, which most scientists had thought would never be detected, had finally been observed, and thoughts turned to catching neutrinos generated in the hidden heart of the Sun.

3.2 NEUTRINOS FROM THE SUN

Solar neutrinos stream through us constantly; they are produced in profusion by thermonuclear reactions in the Sun's core. When hydrogen is burned into helium, thereby making the Sun shine, four protons are united, but two of them have to be turned into two neutrons to make a helium nucleus. Each conversion of a proton into a neutron coincides with the birth of a neutrino, so two neutrinos are created each time a helium nucleus is made.

To make the Sun's energy at its present rate, a hundred trillion, trillion, trillion (ten to the thirty eight) helium nuclei are formed each second (also see Sect. 2.3), and twice that number of neutrinos are released by the Sun every second. Although the Earth intercepts only a small fraction of these, it is still an enormous number. Each second, about 70 billion solar neutrinos fly through every square centimeter of the surface of the Earth facing the Sun, and out through the opposite surface unimpeded.

The Sun therefore bathes our planet with a beam of neutrinos that is as steady as sunlight. And unlike light and other forms of radiation, which lose their identity during their 170-thousand-year journey from the Sun's core, neutrinos pass quickly through the massive body of the Sun with little loss of identity. To neutrinos, the Sun is essentially transparent, so they bring us a unique message from deep within its hidden interior, telling us what is happening in the center of the Sun right now – or to be more exact, about 500 seconds ago – the time required to move at the velocity of light from the center of the Sun to the Earth.

By catching and counting solar neutrinos, we can open the door of the Sun's nuclear furnace and peer inside its energy-generating core. But the feasibility of detecting neutrinos from the Sun depends on exactly what nuclear reactions are involved in making the Sun shine, the internal composition and structure of the Sun, and how our home star has evolved with time.

Fig. 3.2. Solar Neutrino Energies. Neutrinos are produced inside the Sun as a byproduct of fusion reactions in its core, but both the amounts and energies of the neutrinos depend on the element fused and the detailed model of the solar interior. Here we show the neutrino flux predicted by the standard solar model. The largest flux of solar neutrinos is found at low energies; they are produced by the main proton-proton (p–p) reaction in the Sun's core. Less abundant, high-energy neutrinos are produced by a rare side reaction involving boron 8. Broken vertical lines mark the detection thresholds of the gallium, chlorine and neutrino-electron scattering experiments; their detectors are sensitive to neutrinos with energies at the right side of the broken lines. The gallium experiment can detect the low-energy p–p neutrinos, as well as those of higher energy; both the chlorine and neutrino-electron scattering detectors are sensitive to the high-energy boron-8 neutrinos. The neutrino fluxes from continuum sources (p–p and boron 8) are given in the units of number per square centimeter per second per million electron volts (MeV) at one astronomical unit. Neutrinos should also be generated at two specific energies when beryllium-7 captures an electron, and also during a relatively rare proton-electron-proton (pep) reaction; their fluxes are given in number per square centimeter per second. (Adapted from John Bahcall, Neutrino Astrophysics, Cambridge University Press, 1989)

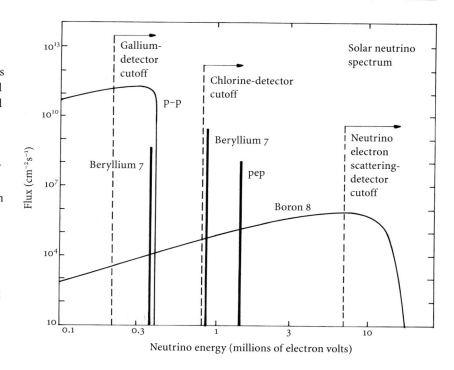

Both the amount and energies of the neutrinos produced inside the Sun depend on the elements being fused (see Fig. 3.2). The great majority of them have relatively low energies and are generated during the head-on collision of two protons. In this proton-proton, or p-p, reaction, one of the protons turns into a neutron, emitting a low-energy neutrino (see Focus 2A); this reaction provides the dominant source of solar neutrinos. However, neutrinos produced in this way have energies of no more than 0.42 million electron volts, or 0.42 MeV.

Scientists have identified several less common neutrino-producing reactions that have not been mentioned earlier because they have no bearing on the energy production of the Sun. The highest-energy neutrinos are produced in one of these rare side reactions. Roughly 15 percent of the time, two types of helium nuclei (helium-3 and helium-4), already produced in the solar core, fuse together to form beryllium-7 and energetic radiation. In rare cases, about once in every 5000 completions of the proton-proton chain, a beryllium-7 nucleus fuses with a proton, forming a radioactive boron-8 nucleus that ultimately decays back into two helium nuclei emitting a neutrino in the process. The boron-8 neutrinos can have energies as great as 15 MeV, or about 36 times more energy than the most energetic proton-proton neutrino. (Willy Fowler, a nuclear astrophysicist at the California Institute of Technology, was one of the first to realize the ramifications of this rare side reaction; he received the Nobel Prize in 1983 for his theoretical and experimental studies of nuclear reactions in the stars and the early Universe.)

More energetic neutrinos find it easier to interact with the material world. However, such encounters are still exceedingly rare, and the Sun produces fewer high-energy neutrinos than low-energy ones. So, if you are in the neutrino-catching business, it's roughly a zero sum game as far as how many neutrinos you might detect at different energies, and it won't be very many of them whatever the energy.

The flux of solar neutrinos expected at the Earth is calculated using large computers that produce an evolving sequence of theoretical models, culminating in the standard solar model that best describes the Sun's luminous output, size and mass at its present age (see Table 2.1). Such calculations have been developed and refined over the past three decades, notably by John Bahcall of the Institute for Advanced Study at Princeton, and his colleagues such as Roger Ulrich at the University of California, Los Angeles, and more recently Marc Pinsonneault at Yale University and Sylvaine Turck-Chièze at Saclay, France.

The computer models always include three basic assumptions:

(1) Energy is generated by hydrogen-burning reactions in the central core of the Sun, and there is no mixing of material between the core and overlying regions. The nuclear reaction rates depend on the density, temperature and composition, as well as coefficients extrapolated from laboratory experiments.
(2) The outward thermal pressure, due to the energy-producing reactions, just balances the inward pressure due to gravity, thereby keeping the Sun from collapsing under its enormous weight.
(3) Energy is transported from the deep interior to the surface via radiation and convection. The great bulk of energy is carried by radiative transport with an opacity determined from atomic physics calculations. (Heavy elements provide a sort of dirt, or opacity, that blocks the flow of radiation, just as a dirty window keeps sunlight from fully illuminating a room.)

One begins with a newly-born Sun having a uniform composition, and an abundance of heavy elements that is observed on the Sun's surface today. The model then imitates the evolution of the Sun to its present age of 4.6 billion years by slowly converting hydrogen into helium within the model core. The central nuclear reactions supply both the radiated luminosity and the local heat (thermal pressure), while also creating neutrinos and producing composition changes in the core. The Sun's current luminosity and size are obtained after 4.6 billion years if about 37 percent of the hydrogen in the core has now been transformed into helium.

Several research groups have developed computer codes for calculating the Sun's evolution. When each calculation is adjusted to the same input parameters, the predicted neutrino fluxes all agree with each other to within a few percent. Any uncertainties in the predictions are attributed to these inputs rather than the computational methods.

Different evolutionary sequences have been computed for slight variations of the original ingredients. An important byproduct of these calculations is an accurate determination of the helium abundance in the Sun. A helium abundance by mass of about 28 percent is obtained; this result relies primarily on the mass, luminosity and age of the Sun. It is consistent with the amount of helium synthesized during the big-bang explosion (roughly 25 percent) that gave rise to the expanding Universe.

To a first approximation, the standard solar model also correctly predicts the frequencies, or periods, of the Sun's surface oscillations (see Sect. 4.3). Thus, the standard solar model meets two independent tests, and most astronomers are convinced that it correctly predicts the flux of solar neutrinos.

Once the neutrino flux has been specified from the standard model, the predictions are extended to specific experiments that detect neutrinos of different energies. The neutrino reaction rate with atoms in these detectors is so slow that a special unit has been invented to specify the experiment-specific flux. This solar neutrino unit, or SNU – pronounced "snew", is equal to one neutrino interaction per second for every trillion, trillion, trillion, or ten to the thirty-six, atoms. And even then, the predictions are only a few SNU per month for even the largest detectors.

3.3 DETECTING ALMOST NOTHING

Of course, it isn't easy to catch the elusive neutrino. The vast majority of neutrinos pass right through matter more easily than light through a window pane, but there is a finite chance that a neutrino will interact with atomic matter. When this slight chance is multiplied by the prodigious quantities of neutrinos flowing from the Sun, we conclude that a few of them will occasionally strike atoms squarely enough to produce a nuclear reaction that signals the presence of an invisible neutrino. The neutrino detector must nevertheless consist of large amounts of material, literally tons of it, to allow interaction with even a tiny fraction of the solar neutrinos and measure their actual numbers.

Unlike a conventional optical telescope, that is placed as high as possible above the obscuring atmosphere, a neutrino telescope is buried beneath a mountain, or deep within the Earth's rocks inside mines. This is to shield it from deceptive signals caused by cosmic interference. There, beneath tons of rock which only the neutrino can penetrate, detectors can unambiguously measure neutrinos from the Sun. (Otherwise, neutrino detectors near the Earth's surface would detect high-energy particles and radiation produced by other energetic particles, called cosmic rays, interacting with the Earth's atmosphere. But neither the primary cosmic rays nor their secondary atmospheric emissions can penetrate thick layers of rock.)

Fig. 3.3. Underground Neutrino Detector. Raymond Davis, who is on the catwalk, and his colleagues built this neutrino trap 1500 meters (about one mile) underground in the Homestake Gold Mine near Lead, South Dakota. It is filled with 100 000 gallons of cleaning fluid. When a high-energy solar neutrino interacts with the nucleus of a chlorine atom in the fluid, radioactive argon is produced, which can be extracted to count the solar neutrinos. Strong signals from energetic cosmic particles are filtered out by placing the detector about 1.5 kilometers below ground; the volume surrounding the tank can be filled with water to provide an additional shield against unwanted signals. This experiment has been run for more than 25 years, always finding fewer neutrinos than expected from the standard solar model. (Courtesy of Brookhaven National Laboratory)

Thus, solar neutrino astronomy involves massive, subterranean detectors that look right through the Earth and observe the Sun at night or day. The first such neutrino telescope, constructed by Raymond Davis, Jr. in 1967, is a 600-ton tank located 1.5 kilometers (about a mile) underground in the Homestake Gold Mine near Lead, South Dakota (see Fig. 3.3). The huge cylindrical tank is filled with 100 000 gallons of cleaning fluid, technically called perchloroethylene or "perc" in the dry-cleaning trade; each molecule of the stain remover consists of two carbon atoms and four chlorine atoms.

Most solar neutrinos pass through the ground and tank unimpeded. Occasionally, however, a neutrino scores a direct hit with the nucleus of a chlorine atom, turning one of its neutrons into a proton, emitting an electron to conserve charge, and transforming the chlorine atom into an atom of radioactive argon. Only neutrinos more energetic than 0.814

million electron volts, or 0.814 MeV, can trigger the nuclear conversion. None of the more abundant proton–proton neutrinos have enough energy to cause this transformation, but the less abundant boron-8 neutrinos are energetic enough to make it take place (also see Fig. 3.2).

The new argon atom rebounds from the encounter with sufficient energy to break free of the perc molecule and enter the surrounding liquid. Because the argon is chemically inert, it can be culled from the liquid by bubbling helium gas through the tank; the number of argon atoms recovered in this way measures the incident flux of solar neutrinos.

Davis and his colleagues have operated their neutrino detector for more than a quarter century, like hunters tending a trap. Every two months they flush the tank with helium, extracting about 24 argon atoms from a tank the size of an Olympic swimming pool; that is a remarkable achievement considering that the tank contains more than a million, trillion, trillion, or ten to the thirty, chlorine atoms. The extracted argon atoms are counted by observing their radioactive decay; every month about half of them capture an electron and change back to chlorine again.

The chlorine experiment detects an average of one neutrino-induced nuclear conversion every 2.5 days, from which one infers an average neutrino flux at the Earth of 2.55 ± 0.25 SNU, where the uncertainties denote one standard deviation. (A standard deviation is a statistical measurement of the uncertainty of a measurement; a definite detection has to be above three standard deviations and preferably above five of them.) In contrast, the standard solar model predicts that it should observe a flux of about 8.0 ± 1.0 SNU. For more than 25 years, the tank full of cleaning fluid has been capturing almost one fourth the expected number of neutrinos (see Fig. 3.4). The discrepancy between the observed and calculated values, which is larger than the uncertainty, is known as the solar neutrino problem.

In 1987, a totally new kind of underground detector began to monitor solar neutrinos; it independently confirmed that the neutrino flux is less

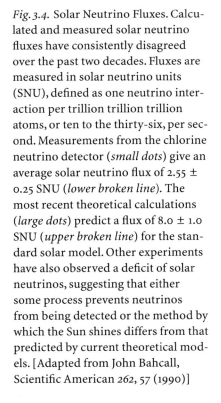

Fig. 3.4. Solar Neutrino Fluxes. Calculated and measured solar neutrino fluxes have consistently disagreed over the past two decades. Fluxes are measured in solar neutrino units (SNU), defined as one neutrino interaction per trillion trillion trillion atoms, or ten to the thirty-six, per second. Measurements from the chlorine neutrino detector (*small dots*) give an average solar neutrino flux of 2.55 ± 0.25 SNU (*lower broken line*). The most recent theoretical calculations (*large dots*) predict a flux of 8.0 ± 1.0 SNU (*upper broken line*) for the standard solar model. Other experiments have also observed a deficit of solar neutrinos, suggesting that either some process prevents neutrinos from being detected or the method by which the Sun shines differs from that predicted by current theoretical models. [Adapted from John Bahcall, *Scientific American 262,* 57 (1990)]

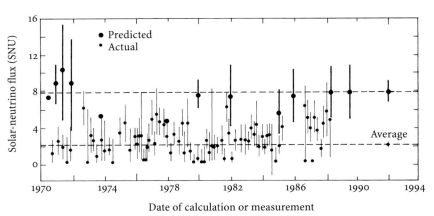

than that predicted by the standard solar model and showed that neutrinos really are coming from the Sun. This second experiment, located in a mine at Kamioka, Japan, and known as Kamiokande II, consists of a 4500-ton tank of pure water. Nearly a thousand light detectors are placed in the tank's walls to measure signals emitted by electrons knocked free from water molecules by passing neutrinos. (Only the central 680 tons of water are used for detecting neutrinos; the outer regions pick up too many confusing signals from the surrounding rock.)

The number of scattered electrons detected by Kamiokande indicates that the flux of high-energy neutrinos is just under half the neutrino flux expected from the standard solar model, or to be exact 0.49 ± 0.06 times the predicted value. Kamiokande II has a much higher threshold than even the chlorine experiment (at 0.814 MeV); an incident neutrino must have an energy of at least 7.5 MeV to produce a detectable recoil electron in water. The dominant source in both experiments is the rare boron-8 neutrinos, so they independently confirm an apparent deficit of high-energy solar neutrinos.

The scattered electrons are detected by the light they emit, and measurements of this light provide important information about the incident neutrinos. When an energetic solar neutrino collides with an electron in the water, the neutrino knocks the electron out of its orbit, pushes it forward in the direction of the incident neutrino, and accelerates the electron to nearly the velocity of light. The electron moves so fast that it sends out shock waves, creating an electromagnetic "sonic boom" much as a jet creates a supersonic boom when it breaks the sound barrier. (In water, the electron moves faster than the light it radiates.) As a result, the electron produces a cone-shaped pulse of light about its path, technically known as Cherenkov radiation, at the precise time that it collides with a neutrino. The axis of the light cone gives the electron's direction, which is the direction from which the neutrino arrived, and the light intensity provides a measure of the electron's recoil energy, as well as the energy of the neutrino.

The electron scattering experiment therefore has a great advantage of recording the direction of the recoil electron. The observed electrons are preferentially scattered along the direction of an imaginary line joining the Earth to the Sun, confirming that the neutrinos are indeed produced by nuclear reactions in the Sun's core. After 1000 days of observation, Yoji Totsuka, speaking for the Kamiokande II collaboration, could therefore report in 1991 that:

> The directional information tells us that neutrinos are coming
> from the Sun, [providing] the first [direct] evidence that the fusion
> processes are taking place in the Sun.[16]

In contrast, the chlorine experiment cannot tell what direction the neutrinos are coming from. However, because the Sun is so close it should dominate the cosmic neutrino input to Earth, and no other known cosmic source could be providing the high-energy neutrinos the chlorine experiment has detected for so long.

The predicted flux of the high-energy boron-8 neutrinos has a large uncertainty, about 37 percent, for it depends strongly on the uncertain temperature at the center of the Sun; these neutrinos vary roughly as the eighteenth power of the core temperature. Scientists have therefore developed a method of detecting the more numerous low-energy neutrinos using gallium, a rare and expensive metal used to make the red lights on hand calculators and other pieces of electronic equipment. Gallium can detect neutrinos produced by the collision of two protons, the most fundamental energy-producing reaction in the Sun. The predicted flux of the proton-proton, or p–p, neutrinos has only a 2 percent uncertainty, primarily because it is relatively insensitive to the core temperature.

When a low-energy neutrino has a rare, head-on collision with the nucleus of a gallium-71 atom (31 protons and 40 neutrons), one of its neutrons is changed into a proton, emitting an electron to conserve charge and producing a nucleus of a germanium-71 atom (32 protons and 39 neutrons). The energy threshold for this nuclear conversion is 0.233 MeV, so it should be sensitive to a sizeable fraction of the p–p neutrino flux, which varies from 0 to 0.422 MeV, as well as the more energetic boron-8 neutrinos (also see Fig. 3.2). Moreover, the germanium produced in this way can be chemically separated from the gallium, and counted by its radioactive decay; every 11 days about half the germanium captures an electron and changes back into gallium. The number of germanium atoms then provides a measure of the flux of incident neutrinos with energies above the 0.233 MeV threshold.

Unfortunately, it takes at least 30 tons of gallium to produce a detectable signal from solar neutrinos, and a ton of gallium costs half a million dollars – a lot more than cleaning fluid! Moreover, a neutrino conversion of a gallium nucleus does not happen very often. Only about one such event is expected to happen each day in 30 tons of gallium. This is serious business, so two international collaborations have spent millions of dollars to bury huge tanks of gallium underground.

In 1990, the Soviet-American Gallium Experiment, or SAGE, began operation in a long tunnel, some 2 kilometers below the summit of Mount Andyrchi in the northern Caucasus. SAGE uses 60 tons of gallium metal kept molten in reactor vessels at about 30 degrees Centigrade. A second multinational experiment, dubbed GALLEX for gallium experiment, started operating in 1991 within the Gran Sasso Underground Laboratory, located 1.4 kilometers below a peak in the Appenine Mountains of Italy. The GALLEX experiment uses 30 tons of gallium in 100 tons of highly concentrated gallium-chloride solution.

Any hope that the solar neutrino problem would be resolved once different detectors were used has now faded away. The two gallium ex-

Table 3.1. Solar neutrino experiments*

Target	Experiment	Threshold energy (MeV)	Measured neutrino flux (SNU)	Predicted neutrino flux (SNU)	Suppression ratio (below prediction)	Neutrino counts
Chlorine 37	Homestake	0.814	2.55 ± 0.25	8.0 ± 1.0	0.32 ± 0.05	750
Neutrino-electron scattering (in water)	Kamio-kande II	7.5			0.49 ± 0.06	380
Gallium 71	SAGE	0.2	73 ± 19	132 ± 7	0.55 ± 0.14	100
Gallium 71	GALLEX	0.2	79 ± 12	132 ± 7	0.60 ± 0.09	136

* Adapted from J. N. Bahcall: Two Solar Neutrino Problems. Physics Letters B *338*, 276–281 (1994). Quoted errors are at the one sigma level. Also see J. N. Bahcall and M. H. Pinsonneault, Astrophysical Journal (Letters) *395*, L119–L122 (1992) and Reviews of Modern Physics *64*, 885–926 (1992).

periments also record a shortage of neutrinos. The total rate predicted for these detectors from the standard solar model is 132 ± 7 SNU, while the GALLEX and SAGE experiments have reported consistently lower measurements of 79 ± 12 SNU and 73 ± 19 SNU, respectively.

Moreover, if you believe the chlorine and the Kamiokande results, which show a deficit of high-energy neutrinos, then the gallium experiments have definitely detected the low-energy p–p neutrinos. This is an important result, for it provides observational confirmation of the theoretical conjecture that hydrogen fusion makes the Sun shine. (The low-energy p–p neutrinos contribute just slightly more than half of the predicted gallium detection rate, or about 74 SNU of the predicted 132 SNU, and the high-energy neutrinos contribute the rest of the predicted amount.)

The results from all four solar neutrino experiments are given in Table 3.1, where they are also compared to theoretical calculations using the standard solar model. The chlorine experiment detects about one fourth of the expected flux of neutrinos, and the neutrino-electron scattering experiment about one half the predicted amount. The two gallium experiments also detect fewer neutrinos than predicted. Taken together, all four experiments seem to have confirmed that the solar neutrinos are missing, and that the solar neutrino problem is real.

3.4 SOLVING THE SOLAR NEUTRINO PROBLEM

After a quarter century of meticulous measurements and calculations, the neutrino count still comes up short! Massive underground detectors always observe fewer neutrinos than theory says they should detect, and the solar neutrino problem is a continuing embarrassment.

But what's all the fuss about anyway? The difference between the standard solar model and observations is only a factor of two or three, or perhaps four. The excitement arises because the stakes are high! Either the Sun does not shine the way we think it ought to, or our basic understanding of neutrinos is in error.

That is, there are two methods of solving the solar neutrino problem. One method is to create a non-standard solar model that modifies our astrophysical description of the Sun and produces the observed number of neutrinos. In the other solution, solar neutrinos are produced in the quantity predicted by the standard solar model, but there is some flaw in our understanding of how subatomic particles behave, requiring new properties for the neutrino.

Is there something amiss with our understanding of the internal operations of the Sun? When the solar neutrino problem first arose, some harried astrophysicists pondered deeply, lost sleep at night, and tore their hair, coming up with all sorts of explanations. Most of them involved a reduction of the temperature deep inside the Sun. If the center of the Sun is about a million degrees cooler than presently thought, the nuclear reactions would run at a slower rate, producing neutrinos in the observed amounts and resolving the dilemma.

A lower central temperature might result from a rapidly rotating core or a strong central magnetic field that would help hold the Sun up against the inward tug of its own gravity, thereby reducing the pressure and temperature at its center; or the temperature might be reduced by fewer heavy elements in the core caused by mixing from the Sun's outer layers. In addition, an imaginary class of subatomic particles, called WIMPs for Weakly Interacting Massive Particles, might lower the core's temperature by extracting thermal energy from it. (If the hypothetical WIMPs exist in large quantities elsewhere in the Universe, they might provide enough invisible mass to eventually halt the expansion of the Universe.) But all these speculative conjectures turned out to be rather wimpy ideas; they require solar models that are inconsistent with the observed surface oscillations of the Sun (also see Sects. 4.3 and 4.4).

Moreover, the relative number of solar neutrinos detected at different energies apparently rules out errors in the standard model. Even a five percent reduction in the Sun's core temperature cannot reconcile all neutrino observations with any reasonable model. If both the Homestake and Kamiokande data are accurate, then a non-standard solar model cannot easily account for them.

FOCUS 3A
Leptons, Quarks and the Electroweak Theory

Electrons, muons and tau particles, along with their corresponding neutrinos, are collectively known as *leptons*, the Greek word for "slender", as they are all significantly less massive than most other elementary particles. The leptons are thought to be fundamental particles that do not consist of anything else; they are therefore amongst the basic building blocks out of which the Universe is constructed.

While the muon is 210 times heavier than an electron, it often behaves like one. But unlike an electron, when a muon is left alone it self-destructs in just 2.2 millionths of a second, decaying into an electron and two neutrinos, one each of both the muon and electron varieties.

Although direct evidence for the tau neutrino has not yet been obtained, its presence is strongly implied by the decay of the tau particle, discovered by Martin Perl and his colleagues in 1974 when using the linear particle accelerator at Stanford University to create the byproducts of electron-positron annihilation. The tau particle is 18 times as heavy as the muon and is extremely ephemeral with a lifetime of only three-tenths of a million-millionth of a second.

We also now know that protons and neutrons are not themselves fundamental, but consist of smaller particles, the quarks, buried deep within them. (The word comes from a phrase of James Joyce – "Three quarks for Muster Mark".) The quarks are bound by the strong force within the protons and neutrons. There are two types of quarks in the proton and neutron, called up and down, and the transmutation of a proton into a neutron involves changing an up quark into a down quark. Every time one quark is changed into another, it produces a neutrino.

Neutrinos occasionally interact with other subatomic particles through a force that is at least 1000 times weaker than the electromagnetic force and 100 000 time more feeble than the strong force. (Electromagnetic force binds electrons to protons, and the strong force holds protons and neutrons together in the nucleus.) The electromagnetic force and the weak force are unified in a single electroweak theory developed by Sheldon Glashow, Abdus Salam and Steven Weinberg; in 1979 they were awarded the Nobel Prize for this feat, even before the discovery of the electroweak forces that were predicted by their theory.

In the standard electroweak theory, the neutrinos are assumed to be completely without mass. There is nevertheless no observational evidence for a mass of zero; the neutrino mass just has to be very small.

By 1990 John Bahcall had teamed up with Hans A. Bethe, who first elucidated the fusion reactions that power the Sun in the late 1930s, arguing that:

> We do not know of any modifications of the astrophysical calculations of the state of the solar interior that could lead to the reconciliation [of the chlorine and neutrino-electron scattering results with theoretical calculations] without requiring new physics for the neutrino.[17]

They showed that any non-standard solar model that is consistent with the Kamiokande data predicts a Homestake chlorine detection of 4 SNU, which is outside the observed range.

The standard solar model, as well as our explanations for how stars shine and evolve, therefore seems to be on solid ground, and the astrophysicists can sleep at night.

So where have all the solar neutrinos gone? The ghostlike neutrinos may be transforming themselves into an invisible form during their journey from the center of the Sun, escaping detection by changing character. The reason we can't see some neutrinos might be because they are hiding in disguise, having undergone metamorphosis into a different form like a caterpillar into a butterfly or moth. After all, other particles, such as the neutron, turn into something else when left alone. Of course, there may not be anything more fundamental for a neutrino to change into, but solar neutrinos could turn into even more evasive types of neutrinos, thereby rendering them invisible.

We haven't mentioned it yet, but the neutrino comes in three types, or flavors, each named after the particle with which it is most likely to interact. All of the neutrinos generated by fusion reactions inside the Sun are electron neutrinos; this is the kind that interacts with electrons and the one predicted by Pauli to explain beta-decay. The other two flavors – the muon neutrino and the tau neutrino – interact with massive, short-lived cousins of the electron, known as muons and tau particles (see Focus 3A).

Neutrinos apparently have an identity crisis! Each type of neutrino is not completely distinct, and the different types interconnect. In the language of quantum mechanics, neutrinos do not occupy a well-defined state; they instead consist of a combination or mixture of states, each with a specific mass. As neutrinos move through space, the mass states come in and out of phase with each other, so the mixture they form changes with time.

In a popular form of this theory, neutrinos can fluctuate from one mass value to another as they propagate through matter, and thereby change from a detectable type of neutrino to an invisible one, like the Cheshire Cat disappearing before it got its head cut off. This neutrino transformation is known as the MSW effect, after Lincoln Wolfenstein of

Carnegie Mellon University, who originated the theory, and the Russian physicists, Stanislaw P. Mikheyev and Aleksei Y. Smirnov, who further developed it.

According to the MSW effect, the electron neutrinos generated at the solar core can change into heavier muon neutrinos on their way out of the Sun. This metamorphosis would happen extremely rarely in the vacuum of space, but might be amplified in the dense interior of the Sun. Interactions between the electron neutrinos and the densely-packed solar electrons can, when the density is just right, alter the mass state of a neutrino travelling out through the Sun, thereby changing it into a muon neutrino. Once formed, the muon neutrino does not change back into an electron neutrino; it travels out into space and remains invisible to current solar neutrino detectors.

This solution to the solar neutrino problem might sound a bit esoteric to those of us who are not specialists in subatomic physics, but informed experts have endorsed it. For instance, in their 1990 paper, Bahcall and Bethe concluded that:

> The explanation of the solar-neutrino problem probably requires physics beyond the standard electroweak model with zero neutrino masses. The experimental results are in excellent agreement with . . . the MSW effect.[18]

There are other researchers who think that different explanations may still be found or that some of the experiments on which Bahcall and Bethe base their conclusion may be incorrect. However, if one accepts the MSW description of neutrino propagation, then the results of all four neutrino experiments can be reconciled with the standard solar model.

In order to change from one form to another, neutrinos must have some substance in the first place. They can only pull off their vanishing act if the neutrino, long thought to be a massless particle, possesses a very small mass. In passing through matter and changing character, the neutrino acquires a miniscule mass in addition to any which it may already have. (That is why Bahcall and Bethe called for an extension of the standard electroweak theory to include neutrinos with mass – also see Focus 3A.) Only one type of neutrino needs to have non-zero mass, and this could be a million times less than the electron's mass, or smaller. Such a tiny mass will probably never be directly measured, and only inferred from the MSW effect.

Confirmation of this effect may have enormous implications. Many scientists think that a grand unified theory, that ties together all nuclear forces, requires non-zero neutrino masses, and the presence of innumerable neutrinos with even a very small mass could affect the entire Universe. So, ghostlike neutrinos from the Sun could take on cosmic importance, providing new insights to subatomic forces and perhaps even the fate of the Universe.

Astronomers and particle physicists are therefore eagerly awaiting results from a new neutrino telescope scheduled for completion in 1996 within a nickel mine near Sudbury, Ontario. The massive detector, located 2 kilometers below ground, will contain 1000 tons of very pure heavy water in a transparent vessel surrounded by 5000 tons of ordinary water and an array of light detectors. (Heavy water contains heavy deuterium atoms in place of light hydrogen ones; Canada has a surplus of heavy water originally produced for use in nuclear power reactors that are no longer in demand.)

Like the chlorine and neutrino-electron scattering experiments, the Sudbury neutrino trap will be sensitive to boron-8 neutrinos and unable to detect less energetic solar neutrinos. But unlike all previous experiments, the new one will detect thousands of events per year; it will collect more neutrinos in just one year than all previous solar neutrino experiments during the past quarter-century. As a result, there will be a great improvement in the uncertainties of previous measurements.

The Sudbury experiment will also be able to see neutrino-electron scattering caused by all three varieties of neutrinos, although scattering events caused by muon and tau neutrinos will be about seven times less common than those produced by electron neutrinos. It should therefore directly test theories in which solar electron neutrinos change into other types of neutrinos, thereby avoiding detection in previous experiments.

Pollard Willows at Sunset. 1888. In this painting, Vincent Van Gogh captures the sulphur yellow and pale gold-lemon color of the Sun. The artist literally worshipped the Sun, which for him represented the source of life and well-being – a well-being that eluded him in his delirium and madness. But even through the iron bars of an insane asylum, Van Gogh never wearied of watching "the Sun rise in its glory". (Courtesy of State Museum Kröller-Müller, Otterlo, The Netherlands)

CHAPTER 4

Taking the Pulse of the Sun

4.1 TRAPPED SOUNDS

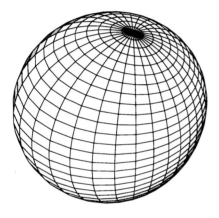

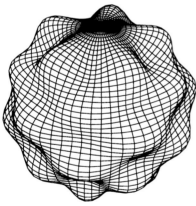

Fig. 4.1. Internal Contortions. The Sun exhibits over a million shapes produced by its internal oscillations. Two of these shapes are illustrated here with an exaggerated amplitude. (Courtesy of Arthur N. Cox and Randall J. Bos, Los Alamos National Laboratory)

The Sun is playing a secret melody, hidden inside itself, that produces a widespread throbbing motion of its surface. The sounds are coursing through the Sun's interior, causing the entire globe, or parts of it, to move in and out, slowly and rhythmically like the regular rise and fall of tides in a bay or of a beating heart (see Fig. 4.1)! This astonishing discovery has been the result of a long, productive interplay between unexpected or unique observations, often with new instrumental techniques, and theoretical explanations or predictions.

In 1960 novel instrumental techniques were used to show that the solar photosphere moves in and out with a rhythmic motion. Such oscillations are imperceptible to the naked eye; the surface moves a hundred-thousandth (0.000 01) of the solar radius. These tiny periodic motions have nevertheless been observed as subtle changes in the wavelength of absorption lines that are formed in the photospheric gas.

When part of the visible surface heaves up toward us, the wavelength of the line formed in that region is shortened, introducing a small blue shift in its spectrum; if the region moves away from us, back toward the solar interior, the wavelength is lengthened, introducing a red shift in the spectrum.

The photospheric motions are inferred by using a computer or photographs to subtract an image of the Sun taken in the long-wavelength side of a stationary (non-moving) absorption line from an image taken in the short-wavelength side of the line. In such a "Dopplergram", an outward motion of a region results in an increase in brightness at that place in the subtracted image, while an inward motion darkens it. (The opposite result occurs if photographic negatives are subtracted.) The Dopplergrams show that bouncing regions move in and out all over the surface of the Sun (see Fig. 4.2).

Such photographic techniques were first employed by Robert B. Leighton and his students, Robert W. Noyes and George W. Simon. They showed that the local vertical motions are not random in their time variation, but instead oscillate with a period of about five minutes. (Curiously, the most intense contractions during childbirth have a similar period of about five minutes, at least during the birth of my children.)

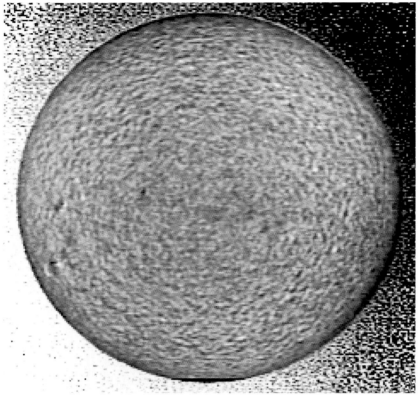

Fig. 4.2. Chaotic Surface. This image of the Sun shows motion toward and away from the observer as light and dark patches, and suggests that the visible solar surface is writhing in chaotic motion. It was obtained using the Doppler effect of a single line (neutral nickel at 6768 Angstroms), thereby measuring the velocity along the line of sight. Large regions of horizontal flow, called the supergranulation, are shown, each about 30 000 kilometers across (the Sun's radius is 700 000 kilometers); they are absent near the disk center where horizontal motions are transverse to the line of sight. Computer analysis of a sequence of such images also reveals the five-minute oscillations caused by sound waves inside the Sun. This image was obtained with the prototype GONG instrument that will soon be used worldwide to obtain nearly continuous recordings of the solar oscillations. (Courtesy of Jack Harvey, Jim Kennedy and John Leibacher, NOAO)

As Leighton announced, at an international conference in 1960, a sequence of Dopplergrams indicates that:

> These vertical motions show a strong oscillatory character, with a period of 296 ± 3 seconds [about five minutes].[19]

The five-minute oscillations had a velocity amplitude of about half a kilometer per second, and covered roughly a third of the solar surface at any given time, localized within numerous patches a few thousand kilometers across. Moreover, the repetitive, vertical oscillations did not seem to be continuous, apparently lasting only a few periods before disappearing.

The initial observations therefore suggested a short-lived and purely local effect that had nothing to do with the Sun as a whole. Perhaps because the overall pattern of the five-minute oscillations seemed chaotic, with each small region oscillating independently of nearby ones, they were attributed to sound waves generated and driven by the small-scale, in and out action of the granulation, somewhat like the repetitive roar of ocean surf crashing into shore.

But the Sun does not resonate with a single pure note – there are millions of them; and the sound waves are not confined to localized regions on the Sun. Careful observations showed that the five-minute oscillations are a global phenomenon, and not correlated with rising granules. The local onset of an oscillation is not associated with the appearance of a granule, and the spatial extent and time coherence of the five-minute oscillations far exceed the size and lifetime of the solar granules. Eventually, the observers demonstrated that the entire Sun is ringing like a bell, with sound waves that resonate within its interior and penetrate to its very core.

In the meantime, F. D. Kahn had shown in 1961 that sound waves can be trapped near the visible solar surface, essentially because the temperature increases both above and below the photosphere. The hotter material acts as a kind of mirror, reflecting sound waves back into the regions they come from. So, the throbbing motions of the solar surface could be caused by sound waves trapped inside the Sun; on striking the surface and rebounding back down, the sound waves cause the gases there to move up and down.

Building upon this result, Leighton and his colleagues concluded in 1962 that :

> The atmosphere may therefore act as a *wave guide* for laterally moving [acoustic] waves, and the observed oscillation may correspond to the lower "cutoff" frequency of the wave guide.[20]

In an extensive observational study of the five-minute oscillations, Edward N. Frazier, at the University of California, Berkeley, then showed, in 1968, that the oscillatory power is concentrated at specific combinations of size and duration, including those where acoustic waves are non-propagating, or evanescent. This finding suggested to him that:

> The well-known five-minute oscillations are primarily standing resonant acoustic [sound] waves … . [They] are not formed directly from the "piston action" of a convective cell [or granule] impinging on the stable photosphere, but rather are formed [in deeper layers] within the convection zone itself.[21]

According to this interpretation, a sound wave resonates within the convection zone, like the plucked string of a guitar, effectively standing in one place and growing in power. This resonance effect is somewhat analogous to repeated pushes on a swing. If the pushes occur at the same point in each swing, they can increase the energy of the motion. In the absence of such a resonance, the perturbations would be haphazard and the effect would eventually fade away. When you regularly move water in a bathtub, the waves similarly grow in size, but when you swish it randomly, the water develops a choppy confusion of small waves.

As suggested in the early 1970s by Roger Ulrich at the University of California, Los Angeles, John Leibacher, then at the Harvard College Observatory, and Robert F. Stein, then at Brandeis University, the convection zone acts as a resonant cavity, or spherical shell. The sound waves are trapped inside this circular waveguide, and can't get out. They therefore go around and around, bouncing repeatedly against the surface like a hamster caught in an exercise wheel, reverberating between the cavity boundaries and driving oscillations in the overlying photosphere.

Many of the sound waves eventually fade away without contributing much to the surface motions. Other sound waves are amplified by repeated reflection, like a swing that is pumped at regular intervals. Such a standing wave hits the surface in the same location, over and over again, pushing the surface in and out at the same places every time it circulates around the Sun (see Fig. 4.3). A sound wave's path inside the Sun then forms a regular sequence of loops, like the lace filigree on a napkin or the hoops in a round rug.

When a sound wave angles up to the photosphere, it strikes it with a glancing blow. None of the waves can escape, because of the sharp decline in density there. (No sound propagates in the vacuum of empty space.) So, an upward-travelling sound wave is turned around and travels back into the Sun, like light reflected from a mirror. This upper turning point is located just beneath the photosphere. Above this level, which is the same for all waves, the sound waves are evanescent and cannot propagate. They instead bounce against the overlying photospheric gases, causing them to rise and fall in a ponderous rhythm.

The inner turning point, or cavity bottom, depends on the increase of sound speed, or wave velocity, with depth. Because the speed of sound is greater in a hotter gas, it increases in the deeper, hotter layers of the Sun. The deeper part of a wave front travelling obliquely into the Sun moves faster than the shallower part, and pulls ahead of it. Gradually, the advancing wave front is refracted, or curved and bent, until the wave is once again headed toward the Sun's surface. (Sound is similarly refracted down into the cool air above a mountain lake; the hotter, higher air bends the sound downward permitting it to travel great distances across the lake's surface.)

So, the increase in sound speed with depth eventually bends a downward-moving sound wave back to the surface of the Sun, and this depth depends upon the period of the wave (also see Fig. 4.3). Long-wavelength sound waves penetrate deep into the Sun before they return, while short waves travel through shallower and cooler layers and bounce off the visible surface more frequently. Moreover, as we shall see, the turn-around depth also depends on physical conditions within the Sun's interior and can be used to determine their radial variation within the Sun.

Each note of the vibrating Sun is similar to the sound wave produced when you tap a crystal glass or strike a doorbell's chime. Something must therefore ring the solar bell, and set the oscillations in motion. And because any oscillation must eventually lose energy and disappear, the solar oscillations must be continually excited.

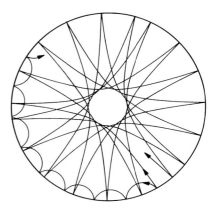

Fig. 4.3. Resonating Sounds. As illustrated by these ray paths, sound waves in the Sun do not travel in straight lines. Instead, the Sun traps them within a spherical shell. The increasing speed of sound toward the center refracts inward-travelling waves outwards. (The effect is analogous to the bending of long radio waves in the Earth's ionosphere or the curvature of light rays in a mirage or within the lens of an eye.) The sharp drop in density at the Sun's visible surface reflects outward-travelling waves back in. How deep a wave penetrates and how far around the Sun it goes before it hits the surface depends on its wavelength. Some waves have just the right length to make an integer (exact) number of bounces before they end up where they began. Such trapped waves interfere constructively with themselves as they circle the Sun, creating resonances that are detectable as ripples, or oscillations, on the solar surface. [Adapted from Douglas Gough and Juri Toomre, Annual Review of Astronomy and Astrophysics 24, 627–684 (1991)]

Sound waves that produce the five-minute oscillations are probably generated by vigorous turbulence in the convection zone; as the hot convective bubbles rise, their motion disturbs the gas they flow through and starts it oscillating, like the hissing noise of a boiling pot of water. Sound is similarly excited in a flute when blown by turbulent air at its mouth; turbulent excitation of sound waves is also responsible for the deafening airport noise of jet aircraft.

4.2 ODE TO THE SUN

The Sun reverberates in millions of ways, creating a resonant symphony of notes, each with its own wavelength and path of propagation through the Sun. About 10 million vibrations are simultaneously excited in the solar interior. Each one is trapped within a well-defined spherical shell, bouncing back and forth as it goes around and around like a Ferris wheel. The combined sound has been compared to a resonating gong in a sandstorm, being repeatedly struck with tiny particles and randomly ringing with an incredible din.

At first, this cacophony seemed hopelessly chaotic and complex, but then a hidden order and regular pattern was discovered in the noise. Theoretical considerations, first developed in 1970 by Roger K. Ulrich of the University of California, Los Angeles, indicate that the trapped sound waves combine and reinforce each other to shake the surface in a regular way. They behave something like cars on a highway that bunch up together, and then disperse as they move along.

To put it another way, standing waves are produced by the constructive interference of sound waves inside the Sun, in much the same way as notes are produced in an organ pipe. As in an organ pipe, which is open at one end, resonance can occur only at particular notes, but an organ pipe resonates in only one dimension; the Sun oscillates in three dimensions and resonates with a greater variety of tones.

The up and down motions observed in the solar photosphere are in reality the combined effect, or superposition, of millions of separate long-lived, global notes or vibrations (Fig. 4.4). Each individual vibration moves the surface of the Sun in and out by only a few tens of meters or less, at speeds of less than ten centimeters per second. When millions of these vibrations are superimposed, they produce surface oscillations with peak values as large as half a kilometer per second, or thousands of times bigger than those of individual vibrations. These large-amplitude combinations are the well-known, five-minute oscillations detected as vertical motions in the photosphere. They grow and decay as individual vibrations go in and out of phase to combine and disperse and then combine again. Groups of birds or schools of fish will similarly gather together, move apart, and then congregate again.

Fig. 4.4. Solar Rhythms. The photosphere moves in and out, rhythmically distorting the shape of the Sun. This heaving motion can be described as the superposition of literally millions of oscillations, including the one shown here. Regions approaching us are blue, and those receding are red. (Courtesy of Jack Harvey, NOAO)

Upon close examination, it is found that five minutes is not the only surface oscillation period. The Sun simultaneously resonates with many different periods from 3 minutes to 1 hour. Moreover, the brief duration and small size of the localized five-minute oscillations is an illusion resulting from the combination of many individual components, each persisting for days or longer and reverberating with spatial extents as small as granules and as large as the Sun itself. Indeed, it is this longevity that enables astronomers to build up a signal and record single notes within the noisy din. (Current observations of global surface oscillations detect velocity amplitudes as low as a few millimeters per second that can persist for nearly one year.)

The vertical motions vary in space and time across the Sun's surface, but only at specific scales. Ulrich, and subsequently Hiroyasu Ando and Yoji Osaki, derived a specific theoretical relation between the permitted sizes and periods, predicting a regular pattern in the apparently random oscillations. For a given size, only certain periods will give a cavity that has the proper depth for a resonant superposition.

The predicted pattern is described in terms of narrow bands, or ridges, of enhanced surface oscillation power when decomposed into a two-dimensional display of size and period, or horizontal wavelength and frequency. The major part of this diagram is covered by modes of the five-minute oscillation. (The spatial extent, or size, of the oscillation is frequently called its horizontal wavelength, while the period is often specified by its reciprocal, the temporal frequency.)

This theoretical conjecture was substantiated in 1975 by the German astronomer Franz-Ludwig Deubner just five years after Ulrich proposed it. Deubner measured the vertical motions in an equatorial strip along the solar disk by recording the Doppler shifts in an absorption line. By using only a few hours of data, taken at high spatial resolution (a few arc-seconds) with a solar telescope on Capri, he was able to show that the oscillatory power is concentrated into narrow ridges in a spatial-temporal display, strikingly confirming the prediction. This meant that the five-minute oscillations are in fact produced by the superposition of numerous sound waves trapped within the Sun.

Similar observational results were obtained in 1977 by Edward J. Rhodes, Jr., working in collaboration with Ulrich and George Simon; just four hours of data obtained with the tower telescope at the National Solar Observatory/Sacramento Peak in New Mexico was enough to substantiate Deubner's remarkable result. Then in 1983 Thomas L. Duvall and Jack W. Harvey, working with the McMath Telescope at the Kitt Peak National Observatory, in Arizona, extended the range of the results, producing spatial-temporal spectra like those shown in Fig. 4.5. They could

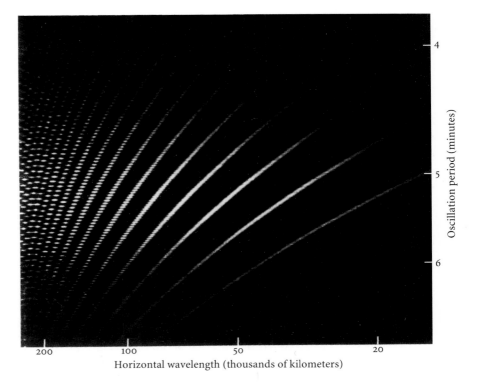

Fig. 4.5. Tones of the Oscillating Sun. Sound waves resonate deep within the Sun, producing surface oscillations with periods near five minutes. Only waves with specific combinations of period and horizontal wavelength resonate within the Sun. The precise combinations are related to the Sun's interior structure; they produce the bright, fine-tuned ridges of greater power shown here. [Courtesy of Jack Harvey, NOAO; also see Tom Duvall and Jack Harvey, Nature *301*, 24–27 (1983)]

Oscillation period (minutes)

Horizontal wavelength (thousands of kilometers)

distinguish more than 20 narrow ridges that cover a wide range of depth within the convection zone, enabling them to improve our understanding of the structure and rotation of the Sun's interior.

Longer-duration observations of the five-minute oscillations have also been obtained; they probe the Sun to large depths and show that it is vibrating to the very core. Such global resonances are detected by measuring Doppler shifts of a single absorption line in light integrated over the entire disk of the Sun. This tends to average out, or cancel, the peaks and troughs of smaller oscillations that occur randomly at many different times and places within the field of view. There is one major problem, however, for the observations require more time then there are hours of daylight.

One solution to this problem, adopted by Eric Fossat and Gérard Grec of the Observatory of Nice, was to travel to the geographic South Pole during the austral summer when the Sun remains at a constant altitude, providing continued observations without interruption by night. (Additional benefits of observing from the South Pole include a cold, transparent atmosphere and the suppression of confusing effects caused by the Earth's rotation; a disadvantage can be long cloudy spells.) Working in collaboration with Martin Pomerantz of the Bartol Research Foundation in Delaware, the French astronomers obtained a continuous 120-hour record of the five-minute oscillations, yielding a power spectrum that showed pairs of well-resolved peaks (Fig. 4.6). This confirmed that the entire Sun rings like a bell with global vibrations that last days and even weeks. Duvall, Harvey and Pomerantz have used long periods of observations taken at the South Pole to determine how rotation varies with latitude as well as depth within the Sun (see Sect. 4.3).

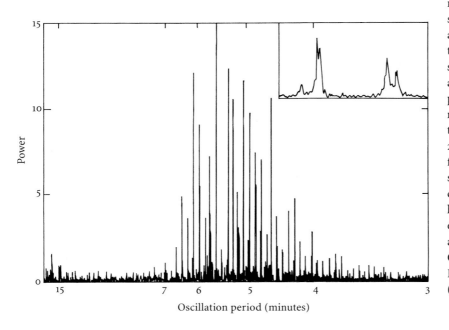

Fig. 4.6. Separating the Modes. The observed power spectrum of low-degree (small l, p-modes) oscillations resulting from a 5-day continuous observation of radial velocity, obtained at the geographic South Pole. It shows that the solar surface is oscillating at specific, well-defined periods denoted along the horizontal axis. Most of the power is concentrated in several very narrow peaks whose largest amplitudes correspond to velocities of only 20 centimeters per second. The inset figure enlarges features within the spectrum. Their paired nature indicated that global modes of low order had been detected, showing that the entire Sun rings like a bell for days and weeks at a time. [Adapted from Gérard Grec, Eric Fossat and Martin Pomerantz, Solar Physics *82*, 55–66 (1983)]

In a complementary approach, pioneered by George R. Isaak and colleagues of the University of Birmingham, hundreds of hours of measurements are obtained in integrated solar light at widely-separated telescopes, such as those at Tenerife, Hawaii and Pic du Midi in the Pyrenees. A spectral analysis of this data also showed discrete features of uniform spacing characteristic of long-lived global oscillations. And in just a few years the result was confirmed by measurements of small fluctuations in the Sun's brightness, obtained with a sensitive radiometer aboard the Solar Maximum Mission satellite.

Thus, in a remarkable sequence of accidental discovery, theoretical prediction, and observational tests, the new science of helioseismology was born.

4.3 LOOKING INSIDE THE SUN

The interior of the Sun is as opaque as a stone wall, and there is no way you can see inside it! But we can illuminate, or sound, the hidden depths of the Sun by recording oscillations at its visible surface.

Some of the oscillations originate just beneath the photosphere; others arise in the Sun's deep interior. We can therefore use them to open a window into the Sun and look at various levels within it. Moreover, the information can be combined to create a picture of the Sun's internal structure, somewhat in the manner of an X-ray CAT scan that probes the inside of a human's brain.

Geophysicists similarly construct models of the Earth's interior by recording earthquakes, or seismic waves, that travel to different depths; this type of investigation is called seismology. Most earthquakes occur just beneath the Earth's surface when massive blocks of rock slip and crunch against one another. The reverberations move out and propagate throughout the terrestrial interior, like ripples spreading out from a disturbance on the surface of a pond.

Solar oscillations probably arise from a persistent, random turbulence in the outer regions of the Sun, and similarly shake it to its very center. In fact, astronomers use the name helioseismology to describe such studies of the Sun's interior; it is a hybrid name combining the Greek words *helios* for the Sun and *seismos* for earthquake.

Seismic waves move in all directions through the Earth and their arrival at various places on the Earth's surface is recorded by seismometers. By combining the arrival times of waves that have travelled through the Earth to various points on the surface, seismologists can pinpoint the origin of the waves and trace their motions through the Earth. This enables them to construct a profile of the Earth's interior in much the way that an ultrasonic scanner can map out the shape of an unborn infant in its mother's womb.

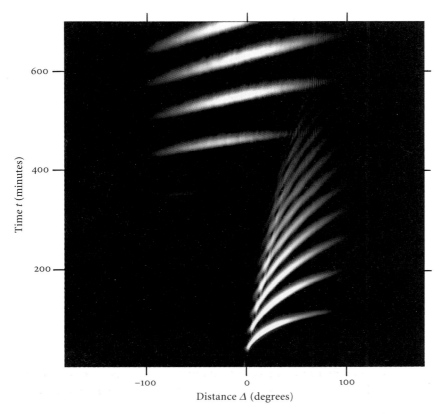

Time *t* (minutes)

Distance *Δ* (degrees)

Fig. 4.7. Time-Distance Helioseismology. By filtering the Sun's five-minute oscillations, one can directly measure travel times and distance for sound waves reverberating in the Sun. The waves take a time, *t,* to travel from the upper reflection point to the bottom of their resonant cavity and back again; distances are measured as an angle between "surface" reflections. Here the lowest curve represents one internal transit of the waves. The curves vertically above the first one correspond to the 2nd, 3rd, etc. reflections from the surface. Different points along a curve correspond to waves penetrating to different depths. [Courtesy of Tom Duvall, also see Nature *362*, 430–432 (1993)]

In analogy with terrestrial seismology, it is now possible to filter the five-minute solar oscillations and isolate waves penetrating to a given depth (Fig. 4.7). That is, one can directly measure the time taken by a sound wave to travel down to the bottom of its cavity and back again, as well as the distance between reflection points.

Of course, unravelling the internal constitution of either the Earth or the Sun is not quite as simple as it might seem. One geophysicist has compared seismology to determining how a piano is constructed by listening to one falling down a flight of stairs. The situation is even worse for the Sun, where the observed five-minute oscillations result from millions of different sound waves, with new ones starting up and old ones dying away all the time, like a bell in a sandstorm.

One therefore has to measure the frequencies of numerous solar oscillations with incredible precision (see Fig. 4.8), and then compare them with theoretical expectations using large computer programs and simplified models of the Sun. Differences between the observed and predicted frequencies can then be used to test and refine the models.

Theoreticians perform this comparison in two ways. The first method, called the "forward problem" is to calculate the frequencies from a set of solar models, each with slightly different interior properties, and see which best fits the data. The second method, in which one works back-

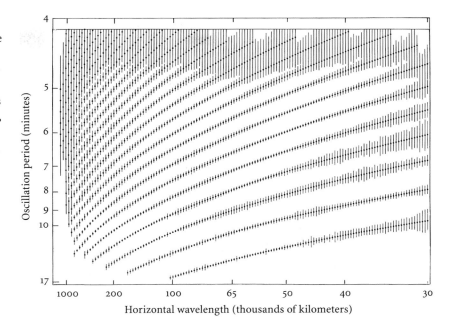

Fig. 4.8. P-Modes. The frequencies, or periods, of sound waves in the Sun are determined with such great precision that the vertical lines shown here represent a thousand times the standard error. Such oscillations are sometimes called p modes because pressure, or p, perturbations provide the main restoring force. [Adapted from Kenneth Libbrecht and M. F. Woodard, *Science* 253, 152–157 (1991)]

ward from the observed frequencies to infer localized properties of the interior, is known as the "inverse problem"; the result is more systematic and less dependent on theoretical assumptions than the forward technique.

And what do such elaborate, drawn-out procedures tell us? The first important conclusion is that calculations using the standard solar model describe the data remarkably well. This agreement means that we can, to a first approximation, adopt the simplifying assumptions of the standard model (see Sect. 3.2); it also reinforces the difficulties already encountered in explaining the low neutrino flux (see Sect. 3.4).

One therefore assumes that the Sun is a gaseous sphere in equilibrium, with an inward rise of temperature that creates the pressure required to balance the weight of overlying layers. Relatively small discrepancies between the computer calculations and observed oscillations are then used to fine-tune the models and thereby establish the detailed radial variation of the Sun's interior temperature, density, sound speed, chemical composition and opacity. (Radial profiles are determined by assuming spherical symmetry, so that all orientations of coordinate axes are equivalent; as described in the next section, rotation breaks this symmetry and produces a second-order effect that can be used to determine the variation of rotation with both depth and latitude.)

This encouraging scenario has but one caveat; the most detailed observations of the five-minute oscillations that are now available refer to only the outer parts of the solar interior, primarily in the convection zone. So, helioseismology most effectively probes the outer half of the Sun, and about 10 percent of its mass. (The convection zone contains only about 2 percent of the mass of the Sun.) In contrast, the solar neutrino flux is most sensitive to an entirely different region of the Sun – the inner energy-generating core.

Solar astronomers have nevertheless already made some startling discoveries using the five-minute oscillations. For instance, there were small but systematic discrepancies present since Deubner's earliest observations, suggesting that the Sun's convective zone extends deeper into the interior than was expected. The theoretical values for the oscillation periods had to be increased by roughly 5 percent, or the oscillation frequency reduced by that amount, to agree with observation. As in the notes produced by an organ pipe, the solar resonances depend on the dimensions of the resonating cavity, and the assumed depth was too shallow. In a "forward" comparison with various models, Douglas Gough concluded, in 1977, that the model depth of the convection zone should be increased by nearly half its previously assumed value to about 30 percent of the Sun's radius.

A more direct determination of the convection-zone depth is derived from measurements of the sound speed, which can be obtained directly from the observed oscillations without the use of solar models. The depth at which a downward-moving sound wave is turned around and bent back toward the surface depends on how quickly the sound speed increases with depth. (Because the speed of sound increases with temperature, it becomes faster in the deeper, hotter layers of the Sun.) By considering a sequence of waves with longer and longer wavelengths, that penetrate deeper and deeper, it is possible to peel away progressively deeper layers of the Sun and establish the radial profile of the sound speed (see Fig. 4.9).

A small but definite change in sound speed marks the lower boundary of the convection zone (also see Fig. 4.9). An accurate inversion of the oscillation data, obtained by Jørgen Christensen-Dalsgaard and his colleagues in 1991, indicates that it is located at a radius of 71.3 percent, or extends to a depth of 28.7 percent, of the radius of the photosphere.

Sound waves are also influenced by the composition of the material they pass through, and can be used to specify the abundance of helium in the convection zone. Comparisons with the observed oscillations yield a helium abundance of between 23 percent and 26 percent by mass, which is consistent with the amount (about 25 percent) that is thought to have been synthesized from primordial hydrogen during the big-bang explosion that gave rise to the expanding Universe.

After corrections for the convection zone depth, small discrepancies remain between the predicted sound-wave periods, or frequencies, and those observed. Refinements in the equation of state, that relates the internal pressure and density, have helped resolve these differences; this equation describes the properties of matter under solar conditions not attainable on Earth.

Fig. 4.9. Sounding the Depths. Observations of the Sun's oscillating surface provide a window into the invisible interior of the Sun, including this determination of the square of the sound speed as a function of fractional radius (zero at the center of the Sun and one at the photosphere). The speed of sound increases at greater depths inside the Sun, and for this reason inward-moving sound waves are refracted outward. The bump at a fractional radius of 0.713 is located at the bottom of the convection zone. Here the thick curve describes observational data, flanked by thin lines representing one standard error. The dashed line is from a theoretical solar model. The large discrepancy between observation and theory at small fractional radius results mainly from the fact that the observed five-minute oscillations do not probe the deep interior very well. [Adapted from Kenneth Libbrecht and M. F. Woodard, *Science 253*, 152–157 (1991)]

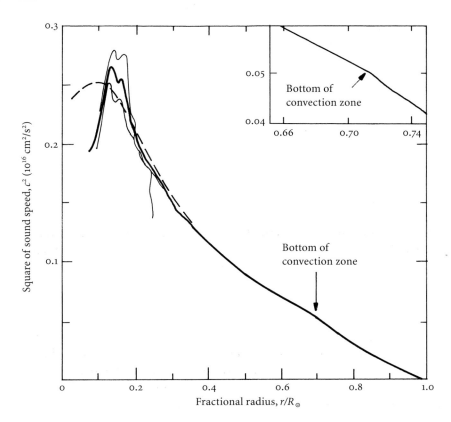

The dominant factor in removing the remaining discrepancies has been corrections to the opacity provided by stable elements heavier than helium. These heavy elements provide the dirt that impedes the outward flow of radiation, like light passing through a muddy car window. Extensive computer calculations of the radiative opacities have resulted in improved solar models and apparently resolved all additional problems. So, theory and observation are now in accord, and astronomers can confidently use the five-minute oscillations to determine second-order effects, including the Sun's internal motions and magnetic structure.

4.4 BREAKING THE SYMMETRY

To a first approximation, the internal structure of the Sun has spherical symmetry, and the frequencies of the five-minute oscillations depend only on radial variations within the spherical shell in which the sound waves propagate. Upon close inspection, however, there are secondary effects that break this symmetry, such as rotation, large-scale magnetic fields and the turbulent motions of convection. These asymmetric forces can produce a fine structure within frequencies that have been calculated under the assumption that they can be neglected.

The most pronounced symmetry-breaking agent is rotation, which can be expressed as a perturbation to the sound waves in a non-rotating star. Waves propagating in the direction of rotation will tend to be carried along by the moving gas, and will move faster than they would in a static medium. (A bird or a jet airplane similarly moves faster when travelling with the wind and takes a shorter time to complete a trip.) The sound wave crests moving with the rotation will therefore appear, to a fixed observer, to move faster and their measured periods will be shorter, or their frequencies higher. Waves propagating against the rotation will be slowed down, with longer periods and lower frequencies.

Thus, rotation imparts a clear signature to the oscillation periods, lengthening them in one direction and shortening them in the other. These opposite effects make the oscillation periods divide, and the observed frequencies are split. Such rotational splitting depends on both depth and latitude within the Sun.

We know that the Sun rotates from east to west, with respect to the stars, at an average period of 25 days at the equator. The solar oscillations have a period of about five minutes, so the rotational splitting is roughly five minutes divided by 25 days, or about one part in ten thousand. The oscillation frequencies have to be measured ten or a hundred times more accurately than this to determine subtle variations in the Sun's rotation, or as accurately as one part in a million.

The rotational properties of the Sun bear directly on its formation and early evolution. In fact, it is a wonder that spinning stars can form at all. The rotation of a shrinking object should speed up as the radius decreases, just as a skater will when her arms are pulled in.

This is because angular momentum, or the product of mass, radius and velocity, is conserved and remains constant, so when the radius becomes smaller the velocity has to increase. Upon collapsing to the Sun's size, an interstellar gas cloud would therefore be rotating faster than the velocity of light, and the outward force of rapid rotation would stop the contraction well before that. So, something must slow the rotation down and carry away excess energy.

Observations show that the surfaces of young stars are indeed rotating much more rapidly than older ones, suggesting that a mature star like the Sun began its life spinning at a much faster rate than it does now. This makes common sense; everything loses energy and slows down as it ages, even we humans.

The outer layers of the Sun could have slowed down significantly over the ages due to the outflow of matter that is magnetically linked to the Sun's surface; indeed, an eternal, magnetized wind continues to flow from the Sun today. Such a wind might carry angular momentum away from the Sun, thereby accounting for its current slow rotation.

If the fast spin of the Sun's youth was retarded by the solar wind, then the Sun ought to be slowed down from the outside in. The interior of the

Sun should therefore be rotating more rapidly than the surface layers, which have shed all the angular momentum with their escaping gas. Instead, the oscillation data indicate that much of the interior rotation is not very different from that seen at the surface, and that the inside is actually spinning at a slightly slower rate than the outside, at least in the equatorial parts of the Sun's convection zone.

The Sun's internal rotation was first determined from the five-minute oscillations using an observing filter squashed into a strip across the solar equator. By isolating sound waves circulating in the equatorial regions (called sectoral acoustic modes) and trapped at different depths, one can infer the radial variation of rotation near the equator, where the Sun's surface rotates at its fastest and the splitting caused by rotation is at its largest. The results of such an inversion, obtained by Tom Duvall and Jack Harvey in 1981, showed a modest inward decline in the equatorial rotation rate of about 15 percent between the surface and about half-way down into the Sun (see Fig. 4.10).

This unexpected finding goes counter to both theory and common sense. Perhaps large-scale circulation within the convective zone mixes and stirs things up, redistributing angular momentum and modifying the effect of solar-wind braking. Perhaps the deep interior retains its faster primordial rotation, and the expected phenomenon will be saved. That is, the core might be rotating fast while the outer parts rotate slowly, with a balance between the angular momentum of the two regions.

In fact, there is marginal evidence for rapid rotation in the innermost energy-generating core of the Sun (also see Fig. 4.10), but the oscillation data are not yet refined enough to accurately determine the core's rotation rate.

Fig. 4.10. Equatorial Rotation Speed with Depth. Accurate measurements of surface oscillations (*inset*) indicate that the Sun is rotating more slowly on the inside than the outside, at least in the outer half of the Sun's equatorial regions. At greater depths, the rotation might speed up, but the data are very uncertain in these regions. The dashed line designates the equatorial rotation frequency at the photosphere. The inset shows the rotation in the outer part of the Sun in greater detail. [Adapted from Thomas Duvall et al., Nature *310*, 22–25 (1984), and courtesy of Sylvain Korzennik (*inset*) Harvard Smithsonian Center for Astrophysics]

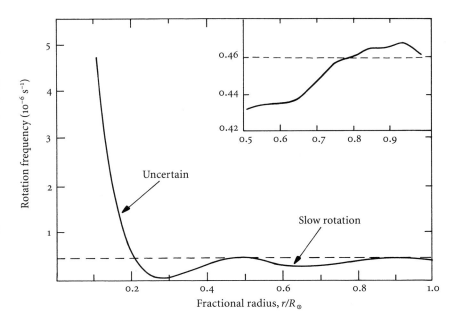

Confirming Einstein's Theory of Gravity

Instead of always tracing out the same ellipse, the orbit of Mercury pivots around the focus occupied by the Sun. That is, the point of closest approach to the Sun, the perihelion, advances by a small amount, 43 arc-seconds per century, beyond that caused by planetary perturbations. Einstein invented an entirely new theory of gravity, called the General Theory of Relativity, that explains this anomalous advance of Mercury's perihelion.

According to this theory, space is distorted and curved in the neighborhood of matter, and this distortion is the cause of gravity. The result is a gravitational effect that departs slightly from Newton's expression, and the planetary orbits are not exactly elliptical. Instead of returning to its starting point to form a closed ellipse after one orbital period, the planet moves slightly ahead in a winding path that can be described as a rotating ellipse. (Because the amount of space curvature produced by the Sun falls off with increasing distance, the perihelion advances for the other planets are much smaller than Mercury's.)

The observed advance of Mercury's perihelion is in almost exact agreement with Einstein's prediction, but this accord depends on the assumption that the Sun is a nearly perfect sphere. If the interior of the Sun is rotating very fast, it will push the equator out further than the poles, so its shape ought to be somewhat oblate rather than perfectly spherical. After all, even the solid Earth is slightly fatter at the middle because of its rotation, and the effect ought to be more pronounced for a rotating gaseous sphere like the Sun. The size of the oblateness, and the amount that it affects gravity, depend on how fast the interior is rotating.

The gravitational influence of the outward bulge, called a quadrupole moment, will provide an added twist to Mercury's orbital motion, shifting its orbit around the Sun by an additional amount and lessening the agreement with Einstein's theory of gravity. Fortunately, the slow rotation of the outer parts of the Sun, which is inferred from the five-minute oscillations, is not enough to produce a substantial asymmetry in its shape, even if the core of the Sun is rapidly rotating. So, we may safely conclude that measurements of Mercury's orbit confirm the predictions of General Relativity under the assumption that the Sun is a nearly perfect sphere.

In fact, the small quadrupole moment inferred from the oscillation data, about one ten millionth rather than exactly zero as Einstein assumed, is consistent with a very small difference between radar measurements of Mercury's orbit and Einstein's prediction. So, the Sun does have an extremely small, middle-aged bulge after all.

When the five-minute oscillations are recorded from the entire solar disk, instead of a strip across the equator, helioseismologists can determine how rotation varies with latitude as well as with depth. The visible surface of the Sun, for example, rotates differently at different latitudes or distances from the equator, with a faster rate at the equator than the poles and a smooth variation in between. The oscillation data indicate that this differential rotation persists within the outer parts of the Sun.

This was first noticed by Duvall, Harvey and Martin Pomerantz in 1986, using long, continuous observations taken at the geographic South Pole, and confirmed in greater detail by several investigators, including Timothy Brown at the High Altitude Observatory and Kenneth Libbrecht at the Big Bear Solar Observatory, using different instruments located at observatories in more pleasant climates. All of the results are in substantial agreement; the differential effect detected at the surface, in which the equator spins faster than the poles, is preserved throughout the convection zone (see Fig. 4.11). Within this zone, there is little variation of rotation with depth, and the inside of the Sun does not rotate any faster than the outside. This relatively slow rotation of the Sun's outer envelope also has important implications for theories of gravity (see Focus 4A).

At greater depths, the interior rotation no longer mimics that of the surface and differential rotation disappears. The internal accord breaks apart just below the base of the convection zone; here the equatorial ro-

Fig. 4.11. Differential Rotation Inside the Sun. Unlike a solid body such as the Earth, the Sun's visible surface (*solid circle*) does not rotate at the same rate at all latitudes, or distances from the equator. This differential rotation persists throughout most of the convection zone where the rotation rate is roughly independent of depth. The entire outer part of the Sun therefore rotates relatively slowly; the equatorial region completes one rotation in about 25 days while it takes 35 days near the poles. In contrast, every point on the solid Earth rotates at the same speed, so a day is 24 hours long independent of latitude. And it's a good thing, for differential rotation would eventually tear the Earth's continents into pieces. (Adapted from helioseismology data obtained by Kenneth Libbrecht at the Big Bear Solar Observatory)

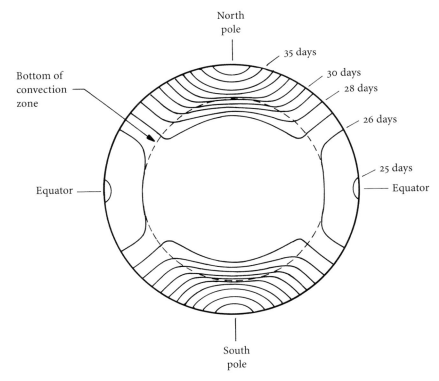

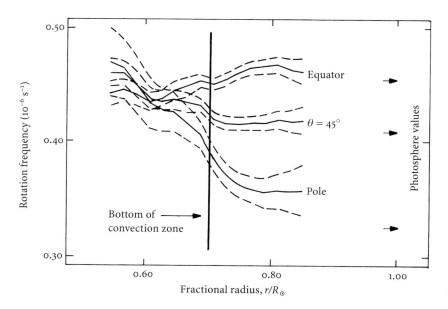

Fig. 4.12. Inside the Spinning Sun. This diagram shows the angular frequency of rotation, plotted as a function of fractional radius, in the equatorial and polar regions, as well as at a latitude of 45 degrees. It shows that the rotation rate deep in the convection zone has a similar latitude variation as that observed at the photosphere, with little variation in depth, but that there is a steep radial gradient in rotation near the base of the convection zone. The polar and equatorial rotation rates become the same nearly half way down into the Sun, where the rotation rate becomes independent of latitude, as in a solid body like the Earth. Dashed lines indicate the observational uncertainties, at one standard deviation, and the photosphere rotation rates from Doppler measurements are indicated by the arrows on the right side. [Adapted from M. J. Thompson, Solar Physics 125, 1–12 (1990) using five-minute oscillation data taken by Ken Libbrecht]

tation slows down going inward and the polar rotation speeds up (see Fig. 4.12). The two rates become equal in the outer part of the radiative interior, where the rotation rate becomes independent of latitude, as it is for a solid body like the Earth. Rotation at deeper levels remains an unknown mystery to be solved in the future.

4.5 BACK TO THE FUTURE

Accurate future measurements of the solar oscillations will provide an extremely detailed picture of the Sun's internal rotation, as well as new information about smaller effects that are superposed on the rotation. These other effects, that might be detected within the convection zone, include its internal, non-spherical temperature structure, large-scale subsurface flows related to possible giant convection cells, and the strength and structure of its magnetic fields. All of these symmetry-breaking agents differ from rotation in that they cannot distinguish between directions that are along or opposed to the direction of rotation.

The major obstacle in obtaining such precise oscillation data is the Earth's rotation, which keeps us from observing the Sun at night. The resultant observational gaps introduce a fundamental uncertainty in the determination of the period of the sound waves, and also create a background noise that hides all but the strongest oscillations. One way to get around this problem is to observe from the South Pole in summer, when the Sun never sets. This has already been accomplished, for about five days at a time, but inclement weather conditions have prohibited longer observations. Astronomers are therefore developing two other methods for obtaining long, uninterrupted views of the Sun – by observing solar oscillations from space, where there can be no night, and by linking Earth-based telescopes in a globe-circling network that follows the Sun as the Earth rotates.

By observing the Sun continuously for long periods of time, one can reduce the background noise in the observed five-minute oscillations, permitting detailed scrutiny of the turbulent convection zone, and also perhaps detecting the low-amplitude signals of longer-period oscillations that penetrate the Sun more deeply. Even global sound waves spend relatively little time in the Sun's energy-generating core, however, so the central regions only modulate stronger effects produced in the overlying regions. This is the reason why very accurate measurements of sound-wave oscillation frequencies have not yet been used to definitively probe rotation at the center of the Sun.

Fortunately, there may be an entirely different class of long-period (hours) resonant oscillations which would have their largest amplitudes near the center of the Sun, and should therefore be much more sensitive to conditions there. They are known as gravity waves, or g-modes, because it is the force of gravity that determines how quickly they rise and fall. (These are totally unrelated to the gravitational radiation predicted by Einstein's General Theory of Relativity.)

Gravity waves occur in regions where there is a stable density difference, or stratification in density. They are produced when a parcel of gas, or fluid, oscillates above and below an equilibrium position, like waves on the surface of the sea. When a high-density parcel moves up into a lower-density region, it is pulled back into place by gravity, and then moves back due to the restoring force of buoyancy. (In contrast, sound waves are restored by pressure, and are sometimes designated p-modes with p for pressure.)

Gravity waves become evanescent, or non-propagating, in regions where the gas is not stably stratified, such as the unstable convection zone. As a result, they are largely confined to the Sun's deep interior where they are the strongest (Fig. 4.13). Gravity waves that manage to reach the visible solar surface have long periods of an hour or more, and are attenuated to low amplitudes as they propagate through the convection zone. They are therefore very difficult to observe.

Yet, there have already been controversial, published reports of detections of these weak, long-period surface oscillations. Future observations of gravity waves, extending over many months and even years, may reliably and precisely establish the properties of the Sun's energy-generating core.

A joint mission of the European Space Agency and NASA, called the Solar and Heliospheric Observatory, or SOHO, is ideally suited for global oscillation measurements. The SOHO spacecraft is scheduled to be launched in 1995, when it will be placed in orbit at the inner Lagrangian

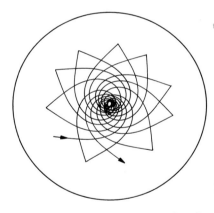

Fig. 4.13. Gravity Waves. Ray paths of gravity waves never reach the visible solar surface and are not reflected by it. Gravity waves are instead turned around inside the Sun, and therefore probe its central depths. Observations of gravity waves might be used to substantiate, or rule out, an increase of the Sun's rotation speed near its center. [Adapted from Douglas Gough and Juri Toomre, Annual Review of Astronomy and Astrophysics 24, 627–684 (1991)]

point of gravitational equilibrium between the Earth and the Sun. This position is continuously in sunlight, permitting uninterrupted observations for years at a time. (The inner Lagrangian point also has a very small line-of-sight, or radial, velocity with respect to the Sun, so instruments aboard SOHO can also make Doppler velocity measurements of surface oscillations with exquisite sensitivity.)

Two full-disk, low-resolution helioseismology experiments aboard SOHO are expected to provide unique, new data on both sound waves (p-modes) and gravity waves (g-modes) that can penetrate the Sun to its very core. These experiments are known as GOLF, an acronym for Global Oscillations at Low Frequencies, and VIRGO, for Variability of solar IRradiance and Gravity Oscillations. They respectively measure variations in the radial velocity, or Doppler effect, and in the Sun's radiative output.

A third SOHO experiment will obtain long, continual, spatially-resolved velocity measurements using a Solar Oscillations Imager, or SOI for short. Precise radial velocities will be observed on spatial scales that cannot be seen from the ground because of the blurring effects of the Earth's atmosphere. This novel space experiment will provide new insight to convective flows and magnetic phenomena that are coded within surface oscillations of relatively small scales and short periods.

In the meantime, a global network of ground-based observatories will make extremely sensitive, uninterrupted observations of the Sun's surface motions, thereby greatly improving the accuracy of more conventional oscillation measurements. This Global Oscillations Network Group, or GONG, consists of six identical, modest telescopes, with 5-centimeter mirrors, of good spatial resolution (a few arc-seconds) spaced roughly equally in longitude, so that at least one site is almost always sunlit and hopefully untroubled by bad weather. (In cloudless skies all you need is three sites separated by 120 degrees in longitude to circle the globe; GONG has six observing stations located at California, Hawaii, Australia, India, the Canary Islands and Chile.) Under normal conditions, each instrument starts observing by itself each morning and shuts down automatically in the evening, without human intervention.

The GONG project will soon be watching, and listening to, the ringing Sun with close scrutiny, taking one picture a minute, every minute around the clock, for at least three years and possibly for a complete 11-year solar activity cycle. It is a truly monumental project, involving immense amounts of data and extensive computations. Each station in the network will produce more than 200 megabytes of data every day. Over the three-year observing run, the raw data will exceed one terabyte and the processed data sets will exceed this several fold; such vast quantities of data are typical of current experiments in high-energy particle physics. (The prefix tera denotes a million million.)

The automated telescopes are making velocity measurements of 65 000 individual points across the surface of the Sun with a precision of better than one meter per second, or better than one part in ten million! When it's all over, the measurement accuracy of the five-minute oscillations will be improved by a factor of ten or more, enabling investigators to explore the hidden solar interior with incredible precision.

In this way, night can be eliminated on Earth and continuous measurements of the oscillating Sun will be obtained, as they can be from space. They will certainly provide new perspectives of the Sun's internal motions, including differential rotation and turbulent convection, and unexpected results are just as likely. After all, if we could predict the future outcome of the new helioseismological observations, there would be little point in doing them. Moreover, future measurements should additionally provide a glimpse of the Sun's internal magnetic field, which has never before been seen.

In fact, the pioneering observations by Leighton and his colleagues showed that the five-minute oscillations have a much lower amplitude inside sunspots than in surrounding regions. (Sunspots are dark, cool islands of intense magnetism found on the visible solar surface; their number and distribution vary with a roughly 11-year activity cycle.) Subsequent observations have shown that the magnetic sunspots remove acoustic energy and absorb as much as half the power of the sound waves that propagate through them.

The intense surface magnetism found in sunspots must extend deep down inside the Sun, where it creates a pressure that modifies the propagation of sound waves, producing a frequency shift that depends on the changing strength of the magnetism. Now, we know that new sunspots apparently drift slowly from the latitudes of about 35 degrees to the Sun's equator during an 11-year cycle, and this surface motion most likely reflects the migration of a deep-seated magnetic field with associated changes in the Sun's internal structure and rotation. These changes may be detected in the future as long-term cyclic variations in the oscillation frequencies. Indeed, one of the principal motivations for helioseismological studies has been a desire to constrain theories for the solar dynamo that produces the magnetic cycle of solar activity, and this therefore brings us to our next topic, the Sun's magnetism.

Grainstack at Sunset near Giverny. 1891. In this painting, Claude Monet captures the faint, reddened sunlight and long shadows at the end of the day. In both the *Grainstack* and *Rouen Cathedral* paintings he portrayed the same object from dawn to dusk, and from summer to winter, to describe the subtle effects of changing light and shadow. (Courtesy of Museum of Fine Arts, Boston, Juliana Cheney Edwards Collection)

CHAPTER 5

A Magnetic Star

5.1 ISLANDS OF INTENSE MAGNETISM

A quick, sideways glimpse of the Sun suggests that its white-hot disk is perfectly round and smooth without a blemish. It's something like glancing at a beautiful woman from a distance, but this rapid, superficial look can be misleading. The Sun's visible surface is often pitted with dark spots, called sunspots, that come and go with lifetimes ranging from hours to months. The ephemeral, dark spots on the Sun indicate that it is an imperfect place of constant turmoil and change. Indeed, everything in the Universe is far from perfect, and everything in it changes.

Sunspots can be seen with the unaided eye through fog or haze, or sometimes at sunrise or sunset, when the Sun's usual brightness is heavily dimmed. (You normally cannot look directly at the Sun without severely damaging your eyes.) Ancient Chinese records show that naked-eye observations of sunspots go back at least 2000 years.

Early in the seventeenth century, Galileo Galilei turned the newly-invented telescope to the Sun. By careful measurements of sunspot positions, Galileo showed that the sunspots are actually on the Sun, and not in front of it. He also used their apparent motion to show that the Sun is spinning in space, and turning around once every 27 days. Galileo noticed that sunspots change in size and shape as they rotate with the Sun, and that they eventually fade and disappear. All of this indicates that the Sun is alive with activity and change, contradicting Aristotle's philosophy of cosmic perfection and immutability.

At about the same time, it was discovered that sunspots near the equator rotate more rapidly than those nearer to the poles; this means that different parts of the Sun's surface rotate about its axis at different speeds, a phenomenon now known as differential rotation. This effect was thoroughly studied two centuries later by Richard C. Carrington, a wealthy English amateur, from his private observatory at Redhill. He showed in 1863 that the Sun's apparent period of rotation increases systematically with latitude from 27 days at the equator to about 30 days halfway toward the poles, or at a latitude of 30 degrees. (Because our planet orbits the Sun in the same direction that the Sun rotates, the rotation rate observed from Earth is about 2 days longer than the Sun's intrinsic rate of spin of about 25 days at the solar equator.)

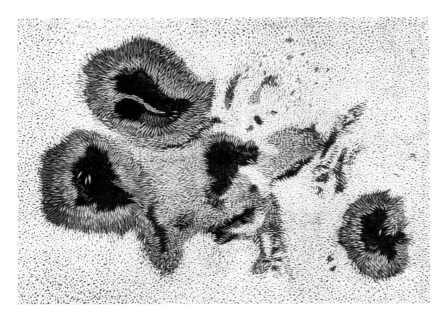

Fig. 5.1. Sunspots. This drawing of sunspots was made more than a century ago, in June, 1861 by J. Nasmyth. (Adapted from Le Ciel, Librairie Hachette: Paris, 1877)

Because they are relatively cool, sunspots appear dark in contrast with their bright surroundings. (A sunspot might have a temperature of 3500 degrees Kelvin, for example, instead of the 6000 degrees Kelvin of adjacent regions.) However, appearances can be deceiving, for the dark sunspots still radiate light. If it were somehow placed alone in space, a sunspot would be about ten times brighter than the full Moon.

Telescopic observations of large sunspots in normal white light indicate a dark center, the umbra, surrounded by a less dark penumbra, both standing in stark contrast to the rather bland and uniform background (Fig. 5.1). Although it looks small in comparison to the solar disk, a large sunspot umbra can surpass the Earth in size. Penumbras are made up of filaments that arch and splay out from the umbra, somewhat like a sea

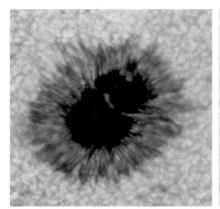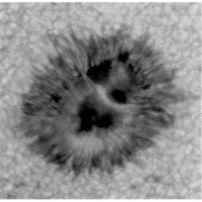

Fig. 5.2. Looking Deep into a Sunspot. Two white-light photographs of different exposure, obtained during exceptional seeing conditions with the McMath Telescope on Kitt Peak. The large sunspot is about 23 000 kilometers (32 arc-seconds) in diameter, or nearly twice as large as the Earth, while the smallest granules seen, at the limiting resolution of the telescope, are about 700 kilometers across. The longer exposure brings out details in the darker umbrae. The umbra of one small sunspot is resolved into an approximately filamentary structure, and a "light bridge" partially spanning the spot has a broader underlying filamentary foundation that completes the crossing. (Courtesy of William Livingston, NOAO)

Fig. 5.3. George Ellery Hale. Twentieth-century solar astronomy was inaugurated when George Ellery Hale developed the first modern solar observatory at Mt. Wilson, California in the early 1900s; he used specialized observing instruments and laboratory experiments to improve our understanding of the Sun. A solar telescope was also installed at his home in Pasadena, together with a large relief of Apollo, the Sun god. Using the spectroheliograph that he invented in 1892, Hale detected the Zeeman effect on the Sun, and used laboratory comparisons to establish the existence of strong magnetic fields in sunspots. In later work, Hale and his colleagues found that the majority of sunspots occur in pairs of opposite magnetic polarity, and that the preceding and following spots have opposite polarity that also reverses sign in the northern and southern hemispheres and from one 11-year activity cycle to the next. (Courtesy of the Archives, California Institute of Technology)

anemone. Fibrils in the filamentary penumbra can nearly extend to the center of the umbra (Fig. 5.2); but dark spots that are sufficiently small, called pores, usually have no surrounding penumbra.

In 1892 George Ellery Hale (Fig. 5.3) invented entirely new ways of looking at sunspots and their surroundings using the solar towers at Mount Wilson, California. Instead of looking at all of the Sun's colors together, Hale devised an instrument, called the spectroheliograph, that would photograph the Sun in just one color or wavelength, without the blinding glare of all the other visible wavelengths. By isolating the bright red emission line of hydrogen atoms, called hydrogen-alpha, as well as the violet calcium K line, Hale opened a new window to the Sun, revealing a changing world of dramatic contrasts.

The spectroheliograph tunes in a thin layer just above the visible photosphere we see with our eye. This layer is called the chromosphere,

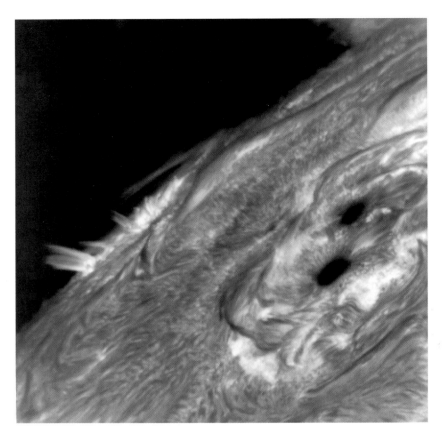

Fig. 5.4. The Red-Faced Sun. At optical wavelengths, solar activity is best viewed by tuning to the red line of atomic hydrogen – the hydrogen alpha line at 6563 Angstroms. Light at this wavelength originates in the chromospheric layers of the Sun, which lie just above the part we see with the eye. An active region, shown in the right half of this image, contains two round, dark sunspots, each about the size of the Earth, and bright plage that marks highly-magnetized regions. Long, dark filaments are held in place by arched magnetic fields. This hydrogen-alpha image of the north-east limb was taken on 26 April 1978 with a 25-cm refractor that has a 4 × 4 arc-minute field of view. (Courtesy of Victor Gaizauskas, obtained at the Ottawa River Solar Observatory, a facility operated by the Herzberg Institute of Astrophysics for the National Research Council of Canada)

from *chromos*, the Greek word for color. It can be detected in red hydrogen-alpha light at a wavelength of 6563 Angstroms or in the violet emission of singly ionized calcium at 3933 and 3967 Angstroms.

Sunspots are dark in the monochromatic hydrogen-alpha photographs, just as they are in white light, but the region surrounding sunspots glows in hydrogen light (Fig. 5.4). This bright neighboring region is called *plage*, from the French word for beach; it is located near dark sunspots within magnetically active regions. (The plages are a chromospheric phenomenon detected in monochromatic light; they are associated with, and often confused with, bright regions of the photosphere, called *faculae* – Latin for little torches, that are seen near the solar limb in white light.) Long, dark filaments also curl across the hydrogen-alpha Sun. They are huge regions of dense, cool gas supported by powerful magnetic forces. Indeed, the Sun's magnetism dominates the hydrogen-alpha world and gives rise to its startling inhomogeneity (Fig. 5.5).

Hale's monochromatic photographs exhibited spiral shapes around sunspots, suggesting to him whirling, vortical motions like those found in the eye of a hurricane or other terrestrial storms. He supposed that the swirling currents of electrified particles would generate a magnetic field, and that the sunspots are therefore magnets. In his own recollections, presented to the National Academy of Sciences in 1913:

> A sun spot, as seen with a telescope or photographed in the ordinary way, does not appear to be a vortex … . But if we photograph the Sun with the red light of hydrogen, we find a very different condition of affairs. In this higher region of the solar atmosphere, first photographed on Mount Wilson in 1908, cyclonic whirls, centering in sun spots, are clearly shown … . Thus we were led to the hypothesis that sun spots are closely analogous to tornadoes or waterspouts in the Earth's atmosphere. If this were true, electrons caught and whirled in the spot vortex should produce a magnetic field. Fortunately, this could be put to a conclusive test through the well-known influence of magnetism on light discovered by Zeeman in 1896.[22]

Of course, the fields and forces of magnetism are invisible. How can we "see" them? As Hale suggested, solar magnetism can be measured by a subtle division and polarization of an atom's spectral lines. This magnetic transformation has been named the Zeeman effect, after the Dutch physicist, Pieter Zeeman, who first noticed it in the terrestrial laboratory.

Fig. 5.5. The Sun in Hydrogen Alpha. This global image of the Sun was taken in the light of hydrogen atoms, emitting at the alpha transition that occurs at a particular red wavelength (6563 Angstroms). It shows small, dark magnetic sunspots, long dark snaking filaments, and bright energetic plages. (Courtesy of the Baikal Astrophysical Observatory, Academy of Sciences, Russia)

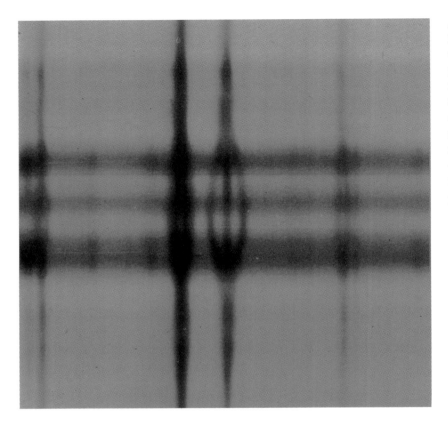

Fig. 5.6. The Zeeman Effect. Spectral lines that are normally at a single wavelength become split into three parts in the presence of a magnetic field. The separation of the components is proportional to the strength of the magnetic field, in this sunspot about 4000 Gauss; their circular polarization, or orientation, indicates the direction, or polarity, of the magnetic field. The Zeeman effect was first applied to the Sun by George Ellery Hale in 1908. (Courtesy of NOAO)

When an atom is placed in a magnetic field, it acts like a tiny compass, adjusting the energy levels of its electrons. If the atomic compass is aligned in the direction of the magnetic field, the electron's energy increases; if it is aligned in the opposite direction, the energy decreases. Since each energy change coincides with a change in the wavelength emitted by that electron, a spectral line emitted at a single wavelength by a randomly oriented collection of atoms becomes a group of three lines of slightly different wavelengths in the presence of a magnetic field (Fig. 5.6). The size of an atom's internal adjustments, and the extent of its spectral division, increase with the strength of the magnetic field.

Furthermore, the light at each of these divided wavelengths has a preferred orientation, or circular polarization, that depends on the direction, or polarity, of the magnetic field. (Lines that are split by a magnetic field that is directed out along the line of sight have right-hand circular polarization, those pointing in the opposite, inward direction have left-hand circular polarization.) So, one can measure the spectral division and use polarized filters to determine both the strength and the direction of the magnetic field using the Zeeman effect.

In 1908 Hale used the 60-foot tower telescope on Mount Wilson to show that spectral lines in the light from sunspots are both polarized and divided, indicating that strong magnetic fields are concentrated within sunspots. By comparing the Zeeman splitting of spectral lines from sunspots with those in laboratory experiments, Hale demonstrated that sunspots have magnetic fields as strong as 3000 Gauss or so, extending over areas larger than the Earth itself. The sunspot's magnetism is thousands of times stronger than the Earth's magnetic field which orients our compasses (about 0.3 Gauss at the equator). The unit of magnetic strength is named after the German mathematician Karl Friedrich Gauss, who showed in 1838 that the Earth's dipolar magnetic field must originate within its interior core.

As sometimes happens, Hale was on the right track but not for all the right reasons. There is no evidence of the giant vortical currents, or storms, envisaged by Hale. The Sun's magnetic fields are instead spawned by an unseen generator deep in the solar interior. When strong, concentrated magnetism breaks through the visible solar surface, a sunspot is formed. Magnetic fields of this strength and size completely dominate the distribution and motion of the charged particles in their vicinity, sometimes giving rise to a surrounding spiral pattern and guiding them into magnetic channels.

Rather than storms, sunspots are relatively calm regions where the concentrated magnetism acts as a filter or valve, choking off the heat and energy (and thus the visible light) flowing outward from the solar interior. The strong magnetic fields in sunspots inhibit the convection currents that usually carry hot material from deeper layers, so less convective heat bubbles up within them. The intense magnetism acts like a refrigerator, keeping sunspots dark and thousands of degrees cooler than the turbulent gas around them. And since it is cooler, a sunspot does not give off as much light as the adjacent material. Indeed, the darkest and coolest parts of sunspots are the sites of the most intense magnetic fields.

At the center of a sunspot, within the umbra, the magnetic field is a few thousand Gauss in strength, and pokes straight out or in from the Sun, like an island of intense magnetism with a fixed direction of north or south polarity. The magnetic field weakens and spreads out in umbrella fashion within the fluted penumbral filaments, where the magnetic field strength is perhaps one thousand Gauss.

Matter flows radially out along the dark penumbral filaments at velocities of a few kilometers per second, an outflow named after John Evershed who first observed it in 1908 from the Kodaikanal Observatory in India. So, the sunspots might not completely block the upward flow of heat. They could redirect it into adjacent areas such as bright plage.

Magnetographs are now used to portray the ever-changing solar magnetism. They consist of an array of tiny detectors that measure the Zeeman effect at different locations on the visible solar disk. Two images are produced, one in each polarization, and the difference of these

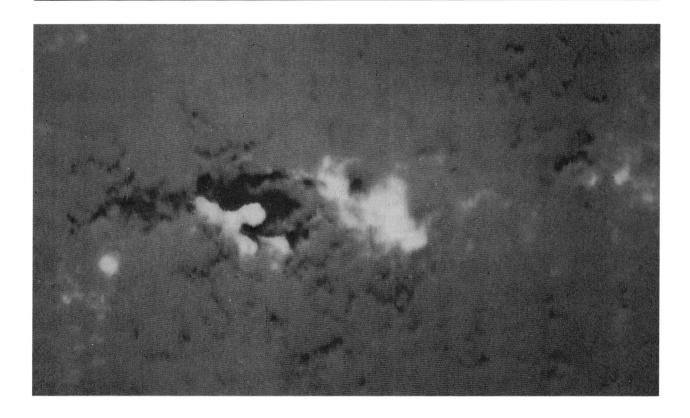

images produces a magnetogram (Fig. 5.7). Strong magnetic fields show up as bright or dark regions, depending on their polarity; weaker ones are less bright or dark.

5.2 BIPOLAR SUNSPOTS, MAGNETIC LOOPS AND ACTIVE REGIONS

Sunspots group together to form the poles of solar magnets, frequently with two principal spots per group (also see Fig. 5.7). One sunspot of each bipolar pair has positive, north, magnetic polarity (or outward-directed magnetism), and its partner has the opposite negative, south, magnetic polarity (inward-directed). The groups of bipolar sunspots are usually oriented roughly parallel to the Sun's equator, in the east-west direction of solar rotation.

The opposite magnetic poles are joined together, like Siamese twins, by magnetic loops that run between them, rising in arches like bridges that connect the bipolar magnetic islands (Fig. 5.8). The magnetic loops can be visualized in terms of the magnetic lines of force, or magnetic field lines, that align compass needles on Earth. The lines of force emerge nearly radially from the sunspot with the positive north polarity and loop through the overlying atmosphere before re-entering the photo-

Fig. 5.7. Twinned Sunspots. In this magnetograph, dark regions denote south polarity and bright ones represent north polarity; it shows bipolar magnetic networks of varying extent. The dark, round sunspot near the center is comparable to the Earth in size. The bipolar structures are joined together in pairs, like Siamese twins; magnetic loops must run between them, stretching up into the overlying atmosphere in an arch. (Courtesy of NOAO)

Fig. 5.8. Loops at the Limb. Magnetic fields channel material into prominent loops that shine in the red light of hydrogen alpha (at 6563 Angstroms). These prominences are detected at the edge, or limb, of the Sun, where they are projected against the dark background. This drawing was made more than 100 years ago; it probably delineates post-flare loops that remain bright for several hours after an eruption of the Sun. (Courtesy of Peter Foukal from Young's *General Astronomy*)

sphere in the spot with negative south polarity, like the lines of force running between the north and south poles of the Earth or a bar magnet. It's as if a powerful magnet, aligned roughly in the east-west direction, was buried deep beneath each sunspot pair.

The magnetized loops that arch above bipolar sunspots can be seen in hydrogen-alpha photographs taken at the solar limb. (The limb is the apparent edge of the Sun's visible surface.) They then appear bright in contrast with the dark background (Fig. 5.9), sometimes extending tens of thousands of kilometers above the solar limb (Fig. 5.10). Hydrogen-alpha photographs also show that limb prominences appear as dark, snaking filaments when projected against the bright photosphere.

The electrically-charged gas that makes up a prominence or filament can hover above the Sun for weeks and months at a time, supported against the downward pull of gravity by the magnetic fields that arch above bipolar regions in the underlying photosphere. The long, thin filaments lie at the tops of magnetic loops, along the magnetic neutral line centered between regions of opposite magnetic polarity. Apparently the gas is held up by numerous magnetic arches, extending in a line, each sagging at the top into a hammock-like shape.

Despite its flamelike appearance when viewed at the solar limb, a prominence is about 100 times cooler and denser than the surrounding material. (As we shall subsequently see, prominences are immersed within the tenuous, million-degree outer atmosphere, or corona, of the Sun.) The magnetic fields that support a prominence or filament also act as a shield and insulate it against hotter surrounding material.

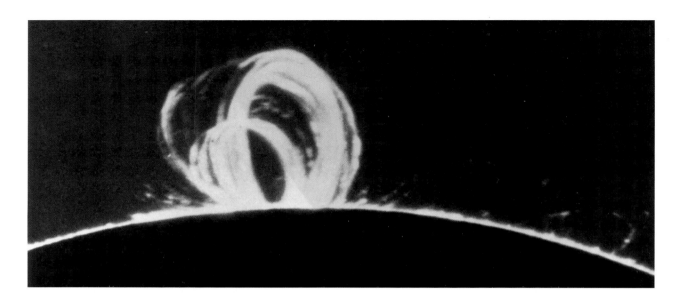

Close comparisons of hydrogen-alpha photographs with magneto-grams have shown that plages are concentrated and heated by the magnetic fields which accompany bipolar magnetic regions. According to one theory, strong oppositely-directed magnetic forces come together and interact or reconnect, to provide the energy that lights up nearby plage. So, both the dark and bright features found in hydrogen-alpha photographs are intimately connected with the Sun's magnetism.

Indeed, most of the phenomena observed at or above the visible solar surface are influenced by, if not controlled by, magnetic fields that guide the motions of the electrically-charged particles found in these regions. A magnetic field presents a barrier to charged particles; they usually cannot cross magnetic field lines and instead move along the lines or spiral around them. Magnetic fields in space provide a conduit for charged particles and guide their motion. In contrast, the motions of un-ionized atoms are not affected by magnetic fields.

The motion of charged particles is described by imagining the invisible magnetic line of force, or magnetic field line, that runs between the bipolar sunspots and outlines the magnetic loop that joins them. This line acts as a guiding center to a moving charged particle. If the particle moves perpendicular to the magnetic field line, the magnetic force acts like a rubber band, pulling the particle back and constraining it to move in small circles about that line.

A charged particle can, however, move freely in the direction of the magnetic line of force. As a result, an electron moves in a spiral path that circles and slides along the magnetic field line. Hot, electrified gas therefore becomes trapped within the closed magnetic loop, moving back and forth within it like a prisoner in an exercise yard.

Fig. 5.9. Loop Prominence. The loop structure of this prominence outlines magnetic fields above sunspots. This photograph was made of an active region at the apparent solar edge, or limb, in the green line of ionized iron, designated Fe XIV. (Courtesy of National Solar Observatory/Sacramento Peak, NOAO)

Fig. 5.10. Arch Prominences. Cool gas, seen in the light of hydrogen alpha, outlines magnetic arches that connect active regions. The prominence material, appearing as a flaming curtain up to 65 000 kilometers above the photosphere, is probably injected into the base of the magnetic loops in the chromosphere. [Courtesy of Big Bear Solar Observatory, Caltech (*top*) and the National Solar Observatory/Sacramento Peak, NOAO (*bottom*)]

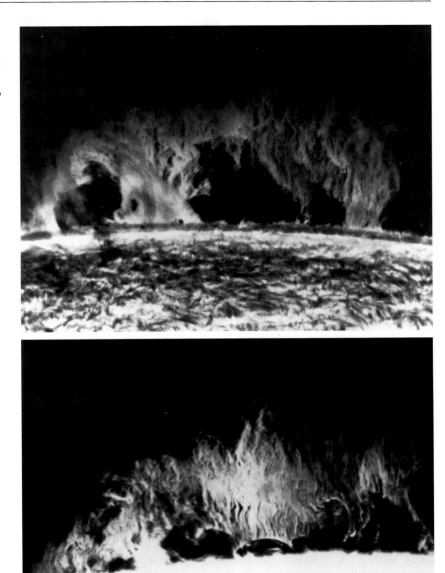

The magnetized realm in, around and above bipolar sunspot groups is a disturbed area called a solar active region. It is a place of concentrated, enhanced magnetic fields, large enough and strong enough to stand out from the magnetically weaker and quieter background atmosphere in images of the whole Sun. An active region is essentially a collection of intense magnetic loops; they together form a magnetic bubble, or magnetic sphere of influence, in which the strong magnetism dominates the motion of charged particles in its vicinity.

Energized material is also concentrated and enhanced within solar active regions where magnetic loops shape, mold and constrain the material and give rise to intense radiation at both visible and invisible wavelengths. Active regions contain relatively cool loops, such as those found in prominences, as well as very hot ones. A hot electrified million-degree gas is, for example, confined within the ubiquitous magnetic loops, where it emits intense radio and X-ray radiation and dominates X-ray images of solar active regions (also see Chap. 6).

Active regions begin their life when magnetic loops emerge from inside the Sun. The magnetic structure of an active region then gradually changes in appearance as new magnetic loops surface within it and its sunspots move and shift about. This results in continued alterations of the form and intensity of their visible and invisible radiation. Eventually, the ephemeral active regions simply disappear. Over the course of weeks to months, their magnetic loops break apart, disintegrate, or submerge back inside the Sun where they came from.

Thus, active regions are never permanent, but instead continually alter their magnetic shape. They are the seat of change and unrest on the Sun! The interacting magnetic forces can, for example, trigger the catastrophic release of magnetic energy stored within active regions, resulting in energetic eruptions, called solar flares (see Chap. 7). Indeed, the continually evolving magnetic structure and intense radiation, as well as the eruptive solar flares, give active regions their name. The whole range of activity varies with the 11-year solar cycle of magnetic activity, which we now focus attention on.

5.3 CYCLES OF MAGNETIC ACTIVITY

The total number of sunspots visible on the Sun varies periodically, from a maximum to a minimum and back to a maximum, in 11 years or so. This sunspot cycle was discovered in the early 1840s by Heinrich Schwabe, an amateur astronomer of Dessau, Germany after seventeen years of observation with a small 5-centimeter (2-inch) telescope, in spite of the fact that other astronomers had uniformly and decidedly asserted that nothing new would be learned from studying sunspots. Instead of varying randomly as had usually been supposed, sunspots periodically rise and fall in total number. As Schwabe stated, after nearly two decades of observations,

[The total number] of sunspots has a period of about 10 years. The future will tell whether this period persists, whether the minimum activity of the Sun in producing spots lasts one or two years and whether this phenomenon takes longer to build up or longer to decline.[23]

Fig. 5.11. Sunspot Cycle. Sunspot positions (*top*) and total sunspot area (*bottom*) have varied in an 11-year cycle for the past 100 years, but this activity cycle varies both in cycle length and amplitude. The top panel shows that the first sunspots of each new cycle appear at about 30 degrees latitude and then spread out to form two bands of active latitudes (one in the north and one in the south) that move toward the equator as the cycle progresses. It also shows how the cycles overlap with spots from a new cycle appearing at high latitudes while spots from the old cycle are still present in the equatorial regions. The total area covered by sunspots (*bottom*) follows a similar 11-year cycle; during each cycle the total area often rises quickly from a minimum to a maximum and then drops back to a minimum at a slower rate. There are large variations in total sunspot area, and in solar activity, during each cycle, and from cycle to cycle. (Courtesy of David Hathaway, NASA/MSFC)

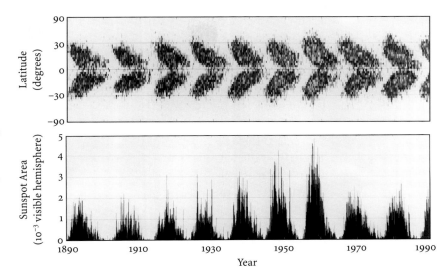

Upon presenting a gold medal to Schwabe, the president of England's Royal Astronomical Society summed up the magnitude of his feat:

> For thirty years never has the Sun exhibited his disk above the horizon of Dessau without being confronted by Schwabe's imperturbable telescope … . This is, I believe, an instance of devoted persistence unsurpassed in the annals of astronomy. The energy of one man has revealed a phenomenon that had eluded even the suspicion of astronomers for 200 years![24]

The sunspot cycle certainly does persist. Other astronomers have now compiled systematic records of the periodic variation in sunspot numbers for hundreds of years (Fig. 5.11).

The sunspot cycle describes a periodic variation in magnetic activity on the Sun, and since most forms of solar activity are magnetic in origin, they also follow an 11-year cycle. The number of active regions, with their energetic radiation and bright magnetized loops, as well as the total number of solar eruptions or flares, also vary from a maximum to a minimum and back to a maximum in about 11 years. However, solar activity does not completely disappear at the minimum in the sunspot cycle.

Magnetograms indicate that there is still plenty of magnetism at sunspot minimum when there are no large spots on the Sun. The magnetism then comes up in a large number of very small regions spread all over the Sun (Fig. 5.12).

The magnetic field averaged over vast areas of the Sun at sunspot minimum is only a few Gauss, but the averaging process conceals a host of fields of small size and large strength. When high-resolution telescopes zoom in to take a close look at smaller and smaller areas, we find evidence of finer and finer magnetic fields with higher and higher field strengths. When the number of sunspots is at a minimum, the magne-

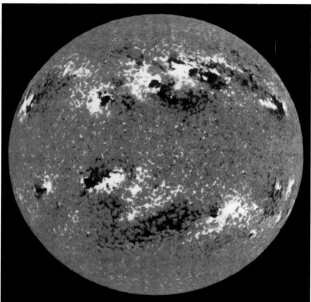

tism is concentrated into these small, intense tubes of magnetic flux that
are typically a thousand Gauss in strength, and separated by wide regions that are relatively field free.

Some of the individual magnetic flux tubes are so small (less than
200 kilometers across) that they cannot be resolved with ground-based
telescopes, and require observations from space to be seen in detail at
visible wavelengths. Although these small, focused pockets of magnetism come and go across the solar surface, they are constantly being re-
generated by internal motions and currents.

So, the magnetic fields are never smoothly distributed across the visible surface of the Sun, either at sunspot minimum or at sunspot maximum. Instead, they are spatially highly concentrated, inhomogeneous,
and everywhere clumped together into intense bundles that cover only a
few percent of the Sun's surface area. At times of maximum activity, the
magnetism is concentrated into active regions within bipolar sunspots
and the magnetic loops that join them.

With the passage of time during the 11-year activity cycle, new sunspots appear closer and closer to the solar equator, while streams of remnant magnetic flux are spread poleward (north or south). In the early
part of the cycle, active regions break out in two belts at about 30 degrees
latitude (see Fig. 5.12). (As on the Earth, latitude is the angular distance
north or south of the equator.) The active-region belts, on each side of
the equator, resemble the backs of two undulating sea serpents that move
parallel to the equator, creating a bipolar sunspot pair each time they
loop through the surface.

◁ *Fig. 5.12*. Magnetic Variations. These magnetograms portray the polarity and surface distribution of the Sun's magnetism. Dark regions have south magnetic polarity, and bright white ones have the opposite north polarity. At times of low activity (*left*, 27 December 1985), there are no large sunspots and tiny magnetic fields of different magnetic polarity can be observed all over the photosphere down to the one arc-second resolution of these images. Notice that dark magnetic elements dominate the north heliographic pole (*top, left*) of the quiet disk, while the bright magnetism at the south pole (*bottom, left*) is mostly of the opposite polarity. This is the Sun's general bipolar field in which magnetic fields loop out of one pole and back into the other. When the Sun is most active, the number of sunspots is at a maximum (*right*, 12 February 1989), and solar magnetism is dominated by large bipolar sunspots oriented in the east-west direction within two parallel bands. By Hale's law of polarities, within a given cycle, the preceding, or westernmost, spots in the northern hemisphere (*top, right*) usually have the same magnetic polarity, while the following (easternmost) spots have the opposite magnetic polarity. In the southern hemisphere (*bottom, right*), the polarities are exactly reversed, forming a mirror image of those in the northern hemisphere. (Courtesy of William Livingston, National Solar Observatory, NOAO)

As the cycle progresses toward sunspot minimum, the active regions at mid-latitudes fade away, and new ones surface in belts that are closer and closer to the equator. The drifting active-region belts, one in each hemisphere, describe a slow, 11-year churning motion that originates deep in the bowels of the Sun and sweeps down across the visible surface, like the jaws of a tightening vice. The belts move inexorably toward the equator until the sunspots come together and disappear at sunspot minimum. Then, out of the destruction, the cycle renews itself once more, and active regions emerge again at mid-latitudes.

This systematic, 11-year drift of sunspots toward the solar equator has been observed for more than a century. It was noticed by Richard C. Carrington in 1858 during his studies of differential rotation, and described by the German astronomer Gustav Spörer, at the turn of the century, in a relation known as Spörer's law. It is graphically represented in a plot of sunspot latitude as a function of time, first drawn by E. Walter Maunder in 1922 and brought up to date in Fig. 5.11. Such an illustration is sometimes called a "butterfly diagram" because of its resemblance to the wings of a butterfly.

Active regions often contain twinned bipolar sunspots that are aligned roughly parallel to the equator, and describe a global pattern of magnetic polarity. All of the sunspot pairs in each active-region belt have the same orientation and polarity alignment, with an exactly opposite arrangement in the two hemispheres (see Fig. 5.12). According to Hale's law of polarity, the leading, or westernmost spots (leading in the sense of rotation) of any sunspot group in the northern belt of active regions have the same magnetic polarity, while the following (easternmost) spots have the opposite magnetic polarity. In the southern hemisphere, the leading and trailing sunspots of any sunspot group also exhibit opposite polarities, but the direction of the bipoles in the southern belt is the reverse of that in the northern one.

Thus, if the leading spot in the northern hemisphere has one magnetic polarity, the leading spot in the southern hemisphere will have the opposite polarity. It's as if men and women always walked down the street in couples, with the men preceding the women on one side of the street and the women leading the men on the other side. Moreover, the couples exactly reverse their orientation at sunspot minimum every 11 years, when the bipolar sunspot magnets flip and turn around, so the leading spots in each hemisphere have opposite magnetic polarities during successive 11-year cycles. (The leading spots resume their original magnetic polarity in approximately 22 years after reversing orientation twice.)

Slow waves of circulation also descend from pole to equator over the course of two 11-year activity cycles, creating zones of fast and slow rotation in the east-west direction. (In technical terms, the pattern of these waves, discovered in 1980 by Robert Howard and Barry L. La Bonte, is that of a torsional oscillator of wave number two.)

Although intense magnetism is distributed within active regions at sunspot maximum, most of this magnetic flux has decayed away, or been dispersed, by the time that sunspot minimum comes around. During sunspot minimum, most of the magnetism is the left-over debris of former active regions, swept and bunched together by convection into a loose network of tiny knots of opposite magnetic polarity (less than 1000 kilometers in size) at the boundaries of supergranules (about 30 000 kilometers across). Some of this remnant magnetism is also transported and shuffled to distant areas on the Sun by deep-seated circulations. (Small magnetic dipoles, called ephemeral active regions, continually bubble up all over the solar surface at sunspot minimum, but apparently make little contribution to the Sun's total magnetic flux.)

5.4 INTERNAL DYNAMO

The Sun's magnetism is constantly being amplified and maintained by internal currents. That is, the hot solar gases, which are good conductors of electricity, move in, out and around the Sun, creating currents that generate and sustain magnetic fields. At the center of the Sun, the electrically charged particles (electrons and protons) are so hot that they conduct an electric current as well as copper does at room temperature, and the entire Sun retains a tremendous current-carrying capacity. Enormous electrical currents of thousands of billions of amperes circulate inside the Sun, producing its awesome magnetism! (For comparison, a 100-watt light bulb carries a current of one ampere.)

The solar interior apparently acts as a dynamo that converts the energy of motion of a conductor into the energy of electric currents and a magnetic field. (The Earth's magnetic field is supposed to be generated by such a dynamo, operating on a much smaller scale within its molten core.) The solar dynamo may be relatively small in scale; it is currently believed to operate only in a thin region at the base of the convection zone. The magnetism is essentially energized by large-scale convective motions, that bring hot gases to the surface from the interior, and by non-uniform, differential rotation in which the Sun revolves faster at the equator than at the poles.

The magnetic fields of the Sun are entrained and "frozen" into the conducting gas whose particles carry the magnetism with them. As they move along with the gas, the embedded magnetic fields are deformed, folded, stretched, twisted and amplified. The mechanical energy of the motion of the charged gas particles is thereby converted into the energy of magnetic fields. This is the basis of the dynamo mechanism, and probably accounts for the continued widespread existence of powerful magnetism throughout the cosmos. (The dynamo mechanism does not ex-

Fig. 5.13. Winding up the Field.   ▷
According to one model, the Sun's highly-conductive, rotating material carries the magnetic field along and winds it up. At the beginning of the sunspot cycle, when the number of sunspots is at a minimum, the magnetic field is the dipolar field seen at the poles of the Sun (*top*). The internal magnetic fields then run south-north between the poles. In this diagram, the Sun's north pole has negative south magnetic polarity with the magnetic field directed inward. Because the equatorial regions rotate at a faster rate than the polar ones, the internal magnetic fields become stretched out and wrapped round the Sun's center (*middle*). The fields are then concentrated and twisted together like ropes. With increasing strength, the submerged ring becomes buoyant and rises to the surface where it penetrates the photosphere, creating magnetic loops and bipolar sunspots (*bottom*). [Adapted from Horace W. Babcock, Astrophysical Journal *133*, 572–587 (1961)]

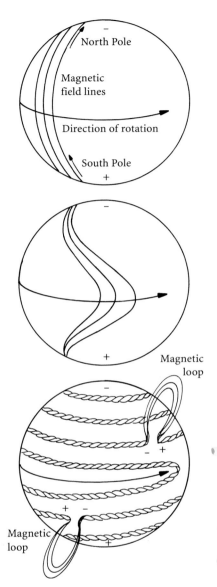

North Pole

Magnetic
field lines

Direction of rotation

South Pole

Magnetic
loop

Magnetic
loop

plain how the cosmic magnetic fields originated, but rather how they are amplified and maintained.)

Although solar physicists do not yet understand what produces the cyclic winding up and relaxation of solar activity, they often extrapolate from a conceptually simple model devised long ago by Horace W. Babcock. His theory begins at sunspot minimum with a global, dipolar magnetic field that runs beneath the surface from south to north, or from pole to pole. Uneven, or differential, rotation shears the electrically conducting gases of the interior, so the entrained magnetic fields get stretched out and squeezed together. The magnetism is eventually coiled and wrapped many times around the inside of the solar globe, ultimately rising to the surface to make active-region belts with their bipolar sunspot pairs (Fig. 5.13).

As Babcock expressed it in 1961:

> Shallow submerged lines of force of an initial, axisymmetric dipolar field are drawn out in longitude by the differential rotation ... . Twisting of the irregular flux strands by the faster shallow layers in low latitudes forms "ropes" with local concentrations that are brought to the surface by magnetic buoyancy to produce *bipolar magnetic regions* with associated sunspots and related activity.[25]

The initial dipolar, or poloidal, magnetic field is twisted into a submerged toroidal (ring-shaped) field running parallel to the equator (east-west). The magnetic fields are amplified when they are bunched together, eventually becoming strong enough to float to the surface and break through it as bipolar sunspots. (The concentrated magnetism is buoyed up by the surrounding gas, just as a piece of wood is subject to buoyant forces when it is immersed in water.) Apparently, the dynamo generates two toroidal magnetic fields (one in the northern hemisphere and one in the southern hemisphere, but oppositely directed) which bubble up at mid- to low-latitudes to spawn the two belts of active regions, symmetrically placed on each side of the equator. Thus, according to Babcock's scenario, we may view the solar cycle as an engine in which differential rotation drives an oscillation between poloidal and toroidal geometries and magnetic polarities.

As the 11-year cycle progresses, the internal magnetic field is wound tighter and tighter by the shearing action of differential rotation, and the two belts of new active regions slowly migrate toward the solar equator. Because the active regions emerge, on the average, with their leading ends slightly twisted toward the equator (see Fig. 5.12), the leading sunspots in the two hemispheres tend to merge and cancel out, or neutralize, each other at the equator. This leaves a surplus of following-polarity magnetism in each hemisphere, north and south, that eventually drifts poleward at sunspot minimum.

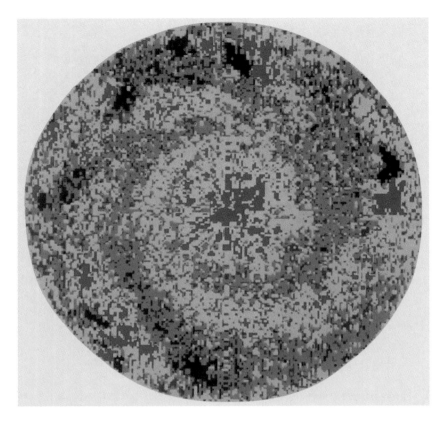

Fig. 5.14. Magnetic Spiral at the Pole. This polar view of the Sun shows poleward streams of magnetism that form a rigidly rotating spiral pattern. Magnetic flux drifts from the equatorial sunspot belt, at the map perimeter, toward the negative-polarity north pole of the Sun, located at the map center. Here red indicates strong negative-polarity photospheric field, dark blue indicates strong positive-polarity field, and yellow or light blue respectively indicate weaker negative-polarity or positive-polarity field. This image was constructed by Yi-Ming Wang, Ana Nash and Neil Sheeley using National Solar Observatory magnetograph data provided by Jack Harvey. (Courtesy of Yi-Ming Wang, Naval Research Laboratory)

Yi-Ming Wang, Ana Nash and Neil Sheeley have recently shown that diffusion and poleward flows sweep remnants of former active regions into streams, each dominated by a single magnetic polarity, that slowly wind their way from the low- and mid-latitude active-region belts to the Sun's poles (Fig. 5.14). By sunspot minimum, when the active regions have largely disintegrated, submerged or annihilated each other, the continued poleward transport of their debris may form a global dipolar field, like the phoenix rising from its ashes. Because the Sun's polar field is created from the following polarity of decaying active regions, they cancel and reverse the overall polarity at sunspot minimum, so the north and south pole switch magnetic direction or polarity. (Thus, it takes two activity cycles, or about 22 years, for the overall magnetic polarity to get back where it started.)

By the time that sunspot minimum cycles around, most of the magnetic flux that emerged in former active regions has been obliterated. It's as if the internal magnetism has been wound so tight that it snaps, like an over-wound watch spring, and no more sunspots can form. Relatively small amounts of magnetism remain, as left-over flux that originated in active regions that are no longer there; it is this flux that has been gradually dispersed over a much wider range of latitudes to form the Sun's polar field. The internal magnetism has then readjusted to a poloidal form, and the magnetic cycle begins again.

Thus, the dynamo theory seems to explain all of the repetitive aspects of the Sun's magnetism, including the periodic variation in the number of sunspots, their cyclic migration toward the equator (the butterfly diagram), the roughly east-west orientation, existence and polarity (Hale's law of polarity) of bipolar sunspot pairs, and the periodic reversal of the overall dipolar field. However, many details of the theory are uncertain or incomplete, and so far no dynamo model has succeeded in accounting for all of the magnetic observations. In fact, some scientists think that a currently-unknown mechanism, other than a dynamo, may eventually account for all the known features of the magnetic cycle.

Helioseismic data indicate, for example, that differential rotation persists throughout the convection zone, and that the rotation does not change significantly with depth within it (see Sect. 4.4). Any dynamo generated in the convection zone would therefore propagate radially outward, rather than creating a symmetric, churning motion from mid-latitudes toward the equator. This suggests that much of the dynamo action may take place deep within the Sun rather than in a shallow layer just below the visible surface, as Babcock originally supposed.

Thus, if it exists, the solar dynamo may operate at the base of the convection zone, where there is a steep radial gradient in rotation, rather than within the convection zone proper. This deep-seated area may periodically wind the poloidal magnetic field present in that region into an azimuthal (toroidal) coil; somehow rising through the overlying convection zone, with its strong differential rotation, to produce sunspots with their cyclic behavior, including the butterfly diagram. But no one has yet observed the magnetic fields deep inside the Sun.

The nature of the Sun's internal magnetic field, as well as its relation to rotation and turbulent motions, may be specified by future helioseismological observations (see Sect. 4.5). They will certainly provide further constraints on the processes that produce cyclic magnetic activity and ultimately couple it to the outer solar atmosphere. In the meantime, the ultimate origin of the activity cycle remains an intriguing unsolved mystery.

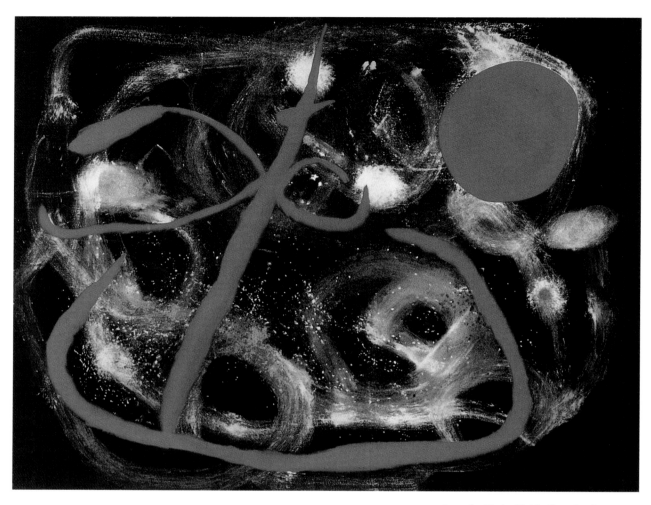

Joy of a Little Girl before the Sun.
1960. In this painting by Joan Miró a
red figure, representing a little girl,
seems to be dancing with the fire-red
Sun and reaching out to the vastness
of space. Some white paint has been
spread upon the black background,
making it appear thin and blue like a
spiral nebula against the black night
sky. (Private Collection)

CHAPTER 6

An Unseen World of Perpetual Change

6.1 THE SUN'S VISIBLE EDGE IS AN ILLUSION

The entire Sun is just a giant incandescent ball of gas that seems to extend forever. The gas is compressed at its center, becoming continuously more tenuous further out. For example, the Sun's outermost visible layer, or photosphere, is about ten thousand times less dense than the air we breathe on Earth. The gases enveloping the photosphere are so rarefied that we can see right through them, just as we see through the Earth's transparent air.

Indeed, we use the term atmosphere for the tenuous outer part of the solar material because it is relatively transparent at visible wavelengths. The photosphere is the lowest, densest level of the solar atmosphere; it is the layer that forms the Sun's visible surface. The unseen outer atmosphere, located above the photosphere and invisible except during solar eclipses or with special instruments, is an energized realm of violent change, extreme temperatures and powerful eruptions that can strongly affect the Earth's environment.

Observing the Sun is like looking into the distance on a foggy day. At a certain distance, the total amount of fog you are looking through mounts up to make an opaque barrier. The fog then becomes so thick and dense that radiation can penetrate no further, and we can only see that far. When looking into the solar atmosphere, you can similarly see through only so much gas. For visible sunlight, this opaque layer is the photosphere, the level of the Sun from which we get our light and heat.

You are looking at the photosphere when you watch the Sun rise in the morning, and continue on its daily journey across the sky. Since you can't see beneath the photosphere, the solar interior remains hidden from view. And you can't use your eyes to see anything in front of the photosphere because we look through the overlying gas.

The yellowy-white light of the photosphere actually comes from a thin, bright shell about 300 kilometers thick, or less than 0.05 percent (0.0005) of the Sun's radius, giving us the thin, sharp-edged Sun we see with the naked eye. (As was described in Sect. 2.4, negative hydrogen ions provide its unusual opacity, even though the photosphere is so diffuse and thin that we would call it a vacuum here on Earth.)

The sharp, visible edge of the Sun is nevertheless an illusion! Being entirely gaseous, the Sun has no solid surface and no permanent visible features. The specification of a "surface" that divides the inside of the Sun from the outside is therefore largely a matter of choice, depending on the wavelength that provides the required perspective.

Just above the photosphere lies a thin layer called the chromosphere, from *chromos*, the Greek word for color. The chromosphere is so faint that it was first observed visually during a total eclipse of the Sun. It became visible a few seconds before and after the eclipse, creating a narrow pink, rose or ruby-colored band at the limb of the Sun. (The solar limb is the apparent edge of the photospheric disk as viewed from the Earth; during a total solar eclipse the Moon just covers the photosphere, and the edge of the Moon coincides with the solar limb.) The spectrum of the chromosphere resembles a reversal of the photospheric spectrum, with bright emission lines at some of the wavelengths where absorption lines appear in the photosphere's light.

The outer boundary of the chromosphere is jagged and irregular, with spiked extensions, called spicules (Fig. 6.1), that shoot up about

Fig. 6.1. Spicules. Rows of dark spicules, or little spikes, are seen in red hydrogen-alpha light. A spicule is a short-lived (minutes) narrow jet of gas spurting out of the solar chromosphere at supersonic speeds to heights as great as 10 000 kilometers. Spicules are a thousand times more dense than the surrounding coronal material, so they are seen in absorption against the bright, chromospheric background. Spicules could provide a magnetic conduit for the outward flow of energy that heats the corona. (Courtesy of National Solar Observatory, NOAO)

10 000 kilometers on average; they rise and fall like choppy waves on a stormy sea or a candle flame in the wind, persisting for only five or ten minutes. (At the polar caps, macrospicules rise a few times higher than other spicules and last a few times longer.) Approximately half a million of the evanescent, flame-like spicules are dancing above the visible Sun's surface at any given moment.

The red color of the chromosphere is supplied by hydrogen atoms emitting at a single red wavelength called hydrogen alpha (at 6563 Angstroms); it is the brightest emission line in the chromosphere. By tuning in the red emission of hydrogen, and using filters to reject all the other wavelengths or colors, we can isolate the light of the chromosphere and produce photographs of it without the blinding glare of the visible photosphere (see Figs. 5.4 and 5.5 of the preceding chapter). In this way, the chromosphere can be observed across the entire disk whenever the Sun is in the sky, rather than just at the edge during a brief, infrequent eclipse.

Still higher, above the chromosphere, is the *corona*, from the Latin word for crown, also momentarily visible to the naked eye when the Sun's bright disk is blocked out, or eclipsed, by our own Moon. (The Moon's shadow sweeps across the Earth at about 1600 kilometers per hour, and the longest total eclipse observed at a fixed point on the ground lasts just under 8 minutes.) During such a solar eclipse we briefly see the Sun's normally invisible corona as a faint, shimmering halo of pearl-white light, extending outward from the lunar silhouette against the blackened sky (Fig. 6.2). The coronal spectrum is unique; its weak colors are crossed by only a few faint emission lines with wavelengths that do not match any of those of the photosphere or chromosphere.

A representative description of the spectacular crown of light, observed during the eclipse of 1842, was provided by Francis Bailey, a stockbroker and enthusiastic amateur astronomer:

> I was astounded by a tremendous burst of applause from the streets below and at the *same moment* was electrified at the sight of one of the most brilliant and splendid phenomena that can be imagined. For at that instant the dark body of the Moon was *suddenly* surrounded with a *corona*, a kind of bright *glory* .... I had anticipated a luminous circle round the Moon during the time of total obscurity ..., but the most remarkable circumstance attending this phenomenon was the appearance of *three large protuberances* apparently emanating from the circumference of the Moon, but evidently forming a portion of the *corona*.[26]

The protuberances that Bailey observed were arches of incandescent gas that loop up into the solar corona and are held there by powerful magnetic forces (also see Chap. 5); astronomers now call them prominences, the French word for protuberances.

Fig. 6.2. Total Eclipse of the Sun. A multiple-exposure photograph of a total eclipse of the Sun. The circular form eclipsing the Sun is the Moon. Because the Moon and the Sun have nearly the same angular extent, the Moon blocks out most of the Sun's light during a total solar eclipse. This photograph was taken by Akira Fujii on 16 February 1980.

Unlike the thin, underlying photosphere and chromosphere, the corona extends throughout the solar system. Indeed, this hot, outermost layer of the Sun's atmosphere is continually expanding into cold interplanetary and interstellar space, creating a solar wind that extends about halfway to the nearest star (see Sect. 6.5). Yet, the coronal electrons are so tenuous and rarefied that a million cubic kilometers of the corona would weigh only about 10 grams.

The intensity of the corona surrounding the Sun is about a millionth of that of the photosphere, so it is no surprise that the corona only becomes visible to the naked eye during a total eclipse. The faint, white coronal halo is, in fact, produced by photospheric light bouncing off electrons set free from their atoms at the hot coronal temperatures. (The reason the corona is so dim is that most of the light from the photosphere passes right through the low-density corona, and only a small fraction of the light strikes the electrons.) However, the corona is only dim by contrast with the bright photosphere; by itself the coronal light could be comparable in brightness to the full Moon.

When we look at the corona during an eclipse, we are seeing patterns of free electrons in the corona made visible because, like motes in a sunbeam, they scatter the light that strikes them. And because these electrons are constrained and molded by magnetic fields, the corona's form varies as the Sun's variable magnetism changes and shifts its shape. (Photospheric light is also scattered off particles of interplanetary dust, producing a component called the F-corona; polarization observations can distinguish between the F-corona (unpolarized) and the intense electron-scattered light (K-corona, polarized)).

White-light pictures, taken when the Moon has blocked out the Sun in total eclipse, show the distribution of the density of electrons in the corona. They are confined within helmet streamers which are peaked like old-fashioned, spiked helmets once fashionable in Europe. (The amount of observed light is proportional to the electron density integrated along

Fig. 6.3. Gossamer Corona. The Sun's corona photographed in white light during the total solar eclipse on 11 July 1991 near sunspot maximum; it extended several solar diameters and had numerous fine rays as well as larger helmet streamers. Japanese astrophotographer Shigemi Numazawa made a total of eight exposures, from 1/15 to 8 seconds long, and combined them with a highly complex darkroom technique to produce this image. Mr. Numazawa is president of the Japan Planetarium Laboratory, located in Niigata, Japan, and is also a renowned astronomical illustrator. (Photograph © 1991 by Shigemi Numazawa)

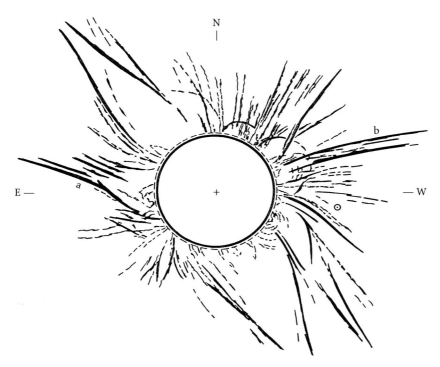

Fig. 6.4. Coronal Structures. Sketch of the main structure of the 11 July 1991 eclipse corona (see Fig. 6.3). The letter "a" marks a sharp streamer edge and "b" unwinding plumes. The capital letters denote the orientation of the hemispheric coordinates. (Courtesy of Serge Koutchmy, Institut d'Astrophysique de Paris, CNRS, and Iraida S. Kim, Sternberg State Astronomical Institute, Moscow State University)

the line of sight.) The electrons are densely concentrated within magnetized loops close to the Sun, creating bubble-like, or arch-like, structures (Figs. 6.3 and 6.4). In the outer part of the corona, far from the Sun, the extensions of these shapes become narrower and surmounted or prolongated by long, straight, tenuous stalks that extend far into interplanetary space (Fig. 6.5).

Helmet streamers are rooted within magnetic loops that sometimes straddle active regions and connect regions of opposite magnetic polarity. Streamers also often rise above long-lived prominences that are commonly embedded in the closed-magnetic loops at their base. These bright, curved, low-lying magnetic loops constrain the densest coronal material close to the photosphere, within one or two solar radii.

Dense, bipolar, magnetically-closed regions of helmet streamers become eventually extended, or stretched out, into open magnetic configurations at large distances from the Sun where the magnetic field weakens. The long, graceful streamers or rays can extend at least ten million kilometers, or 14 solar radii, into interplanetary space, as if some invisible agents were pulling or pushing them out like stretched salt-water taffy. Although the corona is trapped within the underlying magnetically-closed regions, it can flow out along the open magnetic field lines in the long, straight, diverging stalks of the helmet streamers, either by evaporating from the top of their helmet-like bulbs or flowing along their edges. Models suggest that magnetic fields might also connect, or recon-

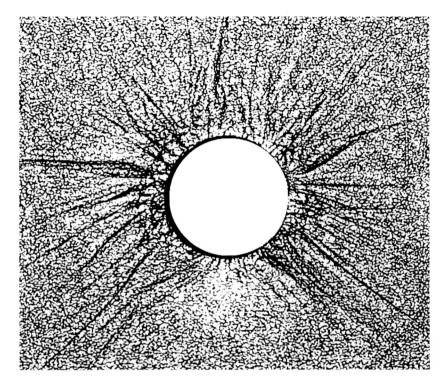

Fig. 6.5. Coronal Rays. This radially-filtered image of the 22 July 1990 eclipse has been digitally processed to enhance low-contrast, fine-scale details of the corona. It is filled with straight, thread-like rays extending as far as six solar radii, or 4 million kilometers, from the center of the Sun. The numerous rays do not point to the exact center of the Sun, so they are not strictly radial. Many of the rays are slightly curved in the inner part of the corona, and some of them apparently begin above the limb of the Sun. North is at the top and east is at the left. (Courtesy of Serge Koutchmy, Institut d'Astrophysique de Paris, CNRS)

nect, and currents could flow along the radially extended parts of streamers where opposite magnetic polarities come into close contact.

The shape of the corona is molded by magnetic forces that vary with the number of sunspots and the amount of solar activity. At sunspot maximum, when magnetic activity is strongest, the streamers are distributed all around the solar limb and presumably all over the Sun (Fig. 6.6). When the number of sunspots is low, the relatively weak magnetic activity is largely confined to the Sun's equatorial regions, where the sunspots and streamers are localized (Fig. 6.7). At this time of reduced solar activity, there is a streamer belt around the coronal equator that extends to form a neutral current sheet, where the magnetic polarity reverses, surrounding the equatorial plane of the Sun (also see Sect. 6.5). The corona is then relatively dim at the poles where faint plumes diverge out into interplanetary space, apparently outlining a weak, global, dipolar magnetic field (also see Fig. 6.6).

Thus, the eclipse observations indicate that the overall shape of the corona changes in synchronism with the 11-year solar activity cycle; near maximum the coronal structures are stretched out in all directions outside the equatorial plane, and near minimum the corona is considerably flattened toward the equatorial regions. The width and radial extension of the body of the streamers is also related to the solar activity cycle. At the time of maximum activity, streamers are smaller and shorter, near

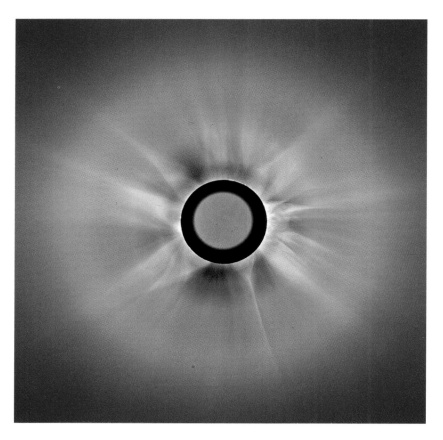

Fig. 6.6. Corona at Sunspot Maximum. The dim, ghostly light of the Sun's thin outer atmosphere, the corona, became visible for a few minutes on 16 February 1980 for the mere 170 seconds of totality during a rare solar eclipse observed from Yellapur, India. This was a time near sunspot maximum, when the sunspots are most numerous; the bright helmet streamers were then distributed about the entire solar limb, resembling the petals on a flower. Wispy streamers of faintly luminous gas reach out 4 million kilometers, or about six solar radii, from the center of the Sun in this photograph. It was taken through a radially graded filter to compensate for the sharp decrease in the electron density and coronal brightness with distance from the Sun. (Courtesy of Johannes Dürst and Antoine Zelenka, Swiss Federal Observatory)

minimum, they are wide and well developed along the equator. (At sunspot maximum, the streamers also deviate from the radial direction near the poles.) It is as if the Sun has only so much magnetism to distribute globally; at maximum activity it disperses the magnetism all over the Sun and weakens its global strength, and at minimum the Sun focuses the magnetism into stronger, more-extended equatorial structures.

The corona can be routinely observed in broad daylight using a special telescope, called the coronagraph, that has a small occulting disk to mask the Sun's face and block out the photosphere's light. The first coronagraph was developed in 1930 by the French astronomer, Bernard Lyot, and soon installed by him at the Pic du Midi observatory in the Pyrenees. As Lyot realized, such observations are limited by the bright sky to high-altitude sites where the thin, dust-free air scatters less sunlight. The higher and cleaner the air, the darker the sky, and the better we can detect the faint corona around the miniature moon in the coronagraph.

The best coronagraph images with the finest detail are obtained from high-flying satellites where almost no air is left and where the daytime sky is truly and starkly black. The clear, nearly continuous, edge-on views of the corona from the Skylab satellite resulted, for example, in the

Fig. 6.7. Corona at Sunspot Minimum. Near the minimum of solar activity, when the number of sunspots is low, bright helmet streamers are concentrated near the solar equator, resembling the wings of a butterfly – as shown in this photograph taken on 30 June 1973 from Loiyengalani, Kenya. The polar regions are then missing bright coronal material, but faint polar plumes can be detected. Near the maximum of the solar activity cycle, when the sunspots are most numerous, the bright helmet streamers are distributed almost randomly about the Sun (see Figs. 6.3 and 6.6). This photograph was taken through a radially graded filter to compensate for the sharp decrease in coronal brightness with distance from the Sun. (Courtesy of Peter A. Gilman and Arthur J. Hundhausen, High Altitude Observatory, National Center for Atmospheric Research)

full realization of extraordinary changes in the corona, including giant Sun-sized, expanding bubbles dubbed coronal transients or coronal mass ejections. Nevertheless, the scattering of residual stray light within all coronagraphs prohibits detection of all but the brightest inner portions of the corona, and they all provide a limited, edge-on view, a flat projection only against the sky.

Because radiation at different wavelengths penetrates to different depths, and therefore originates at different heights, we can focus on particular layers of the Sun's outer atmosphere by tuning through the electromagnetic spectrum, much as adjusting the focus of a pair of binoculars lets you see closer or further away. By combining observations at different wavelengths that penetrate to, and slice through, different layers in the Sun's onion-like atmosphere, we can build up a layered, three-dimensional understanding of the Sun.

What you see on the Sun therefore depends on how you look at it! Indeed, the Sun's complex face changes from wavelength to wavelength, or from layer to layer, as if it wore a different mask for each occasion. When the Sun is observed at invisible X-ray or radio wavelengths, for example, one detects higher, hotter levels in the Sun's atmosphere, permitting a view of the solar corona across the full disk of the Sun. The invisible corona is an energized realm of million-degree temperatures and violent eruptions. As we shall see, such observations have shown us that the apparently serene Sun is continuously changing, seething and writhing in tune with the Sun's magnetism, creating an ever-changing, invisible realm with no permanent features.

6.2 THE MILLION-DEGREE CORONA

During the solar eclipse of August 7, 1869, Charles A. Young and, independently, William Harkness, first found that the spectrum of the solar corona includes a bright, green emission line. This spectral feature appears as a line when the Sun's radiation intensity is displayed as a func-

FOCUS 6A

Taking the Temperature of the Corona

Coronal emission lines detected during a solar eclipse were eventually explained in terms of terrestrially common substances at extreme temperatures of about a million degrees. However, astronomers might have already inferred this high temperature from the great extent of the corona, the large widths of the coronal emission lines, and the absence of Fraunhofer absorption lines in the coronal light.

The Sun's enormous gravity would hold a relatively cool gas close to the photosphere, just as the Earth's gravity binds its atmosphere into a thin shell. But this is inconsistent with the larger extent of the corona seen during an eclipse. A temperature of a million degrees is required to keep the corona extended; the motion of the heated gas supports it against the Sun's gravitational pull.

The detailed character of spectral lines can be used to infer the temperature and motion of the solar gas. The hotter the gas, the faster the motions, and the greater the wavelength shifts from the Doppler effect. These subtle shifts widen the spectral lines, and can be used as a thermometer to measure the corona's temperature. Already in 1941, for example, Bengt Edlén noticed that the observed widths of the iron emission lines in the corona indicate a temperature of about two million degrees.

The corona, which shines by photospheric sunlight scattered from coronal electrons, does not contain the Fraunhofer lines. The Doppler effect causes sunlight reflected from hot, fast electrons to change in wavelength, causing the dark Fraunhofer absorption lines to disappear; they become smeared out until they are invisible.

The very high coronal temperatures might have indeed been inferred from their absence. In contrast, atoms in the photosphere, at a "mere" 6000 degrees Kelvin or so, aren't moving fast enough to wash out these spectral features.

Astronomers were probably reluctant to accept all of this evidence because the emission lines might be due to unusual substances, and also because they did not expect heat to flow from the cool photosphere into the surrounding hotter corona.

tion of wavelength; the specific wavelength of each line fingerprints the atom or ion from which it originated. However, the green coronal emission line seen in the eclipse spectra of the solar corona could not at first be identified with any known element, so astronomers initially concluded that the corona contains a previously unknown substance, which they named coronium.

Belief in the new element coronium lingered for many years, until it became obvious that there was no place for it in the atomic periodic table, and it must therefore be, not an unknown element, but a known element in an unusual state. The solution to the coronium puzzle nevertheless eluded astronomers for seventy years. During this interval, eclipse observations revealed at least ten coronal emission lines, none of which had been observed to come from terrestrial substances. Then in 1941 the Swedish astronomer Bengt Edlén, following a suggestion by Germany's Walter Grotrian, showed that the coronal emission lines are emitted by ordinary elements such as iron, calcium and nickel but from atoms deprived of 10 to 15 electrons. The prominent green emission line (at 5303 Angstroms) is, for example, due to iron atoms that have been stripped of half of their 26 electrons (labelled Fe XIV).

Edlén realized that the highly ionized atoms could only be missing so many electrons if the coronal gas was unexpectedly hot, a million degrees or more (also see Focus 6A)! At these high temperatures, many electrons are set free from atoms and move off at high speeds, leaving the ions behind. (The positively charged ions are atoms stripped of some electrons; these ions move more slowly than the free electrons because the ions have a greater mass). The free coronal electrons are moving so fast that they can easily knock off other electrons when they strike an atom or ion. At a million degrees, only the heavy atoms, such as iron, are able to hold on to any of their orbiting electrons, and even they can only keep some in their grasp.

The coronal lines cannot be observed in the terrestrial laboratory and are hence designated as "forbidden" lines. According to quantum theory, electrons that are still attached to an ion can be rearranged into certain long-lived orbits when the ion is excited, and the electrons emit the forbidden lines when they eventually move out of these excited orbits. However, even in the best vacuum on Earth, frequent collisions knock the electrons out of these orbits before they have a chance to emit the forbidden lines.

The free electrons in the corona, which are not attached to anything, scatter sunlight from the photosphere and illuminate the corona. The electrons bend small amounts of the sunlight into our line of vision, just as tiny dust particles illuminate a sunbeam in your room and air molecules scatter sunlight to make the sky blue. Yet, the vacuous corona scatters so little sunlight that it is a million times fainter than the photosphere at visible wavelengths. And because the corona is so tenuous, most of the photospheric light passes right through the corona without being scattered.

Although the electrified coronal particles move about at great speed, there are so few of them that the total energy in the corona is quite low. Only about a millionth of the Sun's total energy output is required to heat the corona. And even though the free electrons are extremely hot, they are so scarce and widely separated that an astronaut or a satellite will not burn up when immersed in the rarefied corona. If the corona were as dense as the underlying regions, at a temperature of a million degrees it would contain enough energy to vaporize the Earth.

There are about a billion electrons and protons per cubic centimeter (10^9 cm^{-3}) at the base of the corona, which sounds like a lot, but the low corona is one hundred times more tenuous than the chromosphere (10^{11} cm^{-3}) and an additional million times more rarefied than the photosphere (10^{17} cm^{-3}). This average drop in density with increasing distance from the Sun is matched by an overall increase in temperature from about 6000 degrees Kelvin in the photosphere to almost 10 000 degrees in the chromosphere and a million degrees at the base of the corona (Fig. 6.8). The linkage between the chromosphere and the corona occurs in a very thin transition region, less than 100 kilometers thick, where both the density and temperature change abruptly; the density decreases as the temperature increases in such a way to keep the gas pressure spatially constant (also see Fig. 6.8). The corona then thins out and slowly cools with increasing distance from the Sun, reaching a density of only about 5 electrons and 5 protons per cubic centimeter at the Earth's orbit, where the coronal temperature has slightly decreased to about 100 000 degrees Kelvin.

Fig. 6.8. Transition Region. Temperature and gas mass density are plotted on logarithmic scales as a function of height in kilometers for average quiet regions of the Sun. The temperature decreases from values near 6000 degrees Kelvin at the visible photosphere to a minimum value of roughly 4400 degrees Kelvin at the base of the chromosphere about 500 kilometers higher in the atmosphere. Thereafter, the temperature increases, slowly at first, then extremely rapidly in the narrow transition region between the chromosphere and corona where the temperature surges upwards by a factor of about one hundred from about 10 000 degrees Kelvin to about one million degrees Kelvin. The reason for this abrupt increase in temperature within the transition region, which is less than 100 kilometers thick, is still not fully understood. (Courtesy of Eugene Avrett, Smithsonian Astrophysical Observatory)

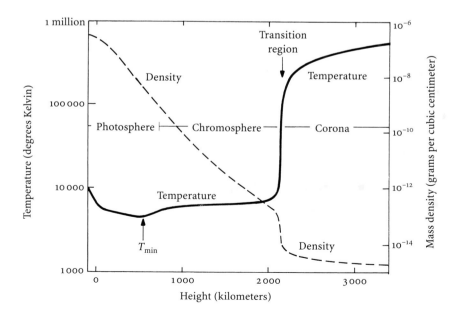

The searing temperatures in the corona have been substantiated by its radio and X-ray radiation. When a high-speed electron passes near an ion, their opposite charges interact electrically and the electron's path is bent by the pull of the ion's charge; the heavy ion shows only a small deflection from its path but the lighter electron accelerates and completely changes its direction, emitting radiation at radio and X-ray wavelengths in the process. Astronomers have named it for the German word *bremsstrahlung*, meaning braking radiation. The strength of the radio and X-ray bremsstrahlung has been used as a thermometer, confirming the corona's million-degree temperature. (This radiation is also known as thermal radiation since it depends on the random thermal motion of the hot electrons, and the electron-ion interaction is termed free-free, because the electrons remain free of atomic bonds before and after the interaction.)

In both the radio and X-ray regions of the electromagnetic spectrum, the hot corona can be seen all across the Sun's face with high spatial and temporal resolution (Fig. 6.9). This is because the Sun's visible photosphere, being so much cooler, produces negligible amounts of radio or X-ray radiation. Indeed, the corona is so hot that it emits most of its energy as X-rays, which are high-energy photons. Although less energy is emitted at radio wavelengths, they can be used to probe different levels within the corona, while also specifying the strength and structure of its magnetic field. Yet, despite today's ability to obtain face-on views of the corona from both ground-based (radio) and space-borne (X-ray) telescopes, the exact mechanisms for heating the solar corona and sustaining it still remain a mystery.

The visible solar surface, or photosphere, is closer to the Sun's core than the million-degree corona but is several hundred times cooler, and this comes as a big surprise. The essential paradox is that energy should not flow from the cooler photosphere to the hotter corona anymore than water should flow uphill. When you sit far away from a fire, for example, it warms you less.

The temperature of the corona is just not supposed to be so much higher than that of the atmosphere immediately below it. It violates common sense, as well as the second law of thermodynamics, which holds that heat cannot be continuously transferred from a cooler body to a warmer one without doing work. This unexpected aspect of the corona has baffled scientists for decades, and they are still trying to explain it.

We know that the visible photospheric sunlight cannot resolve the heating paradox. Heat from the photosphere does not go into the corona, it goes through the corona. There is so little material in the corona that it is transparent to almost all of the photospheric radiation. Sunlight therefore passes right through the corona without depositing substantial quantities of energy in it, travelling out to warm the Earth and to also cool the photosphere off in spite of the hot corona that envelops it. (Fur-

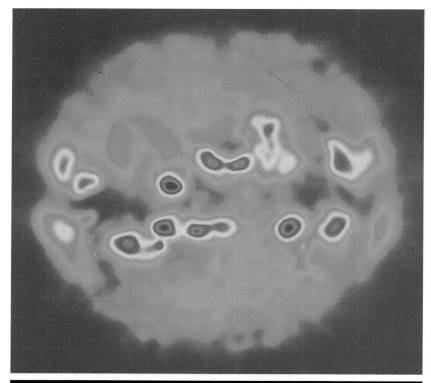

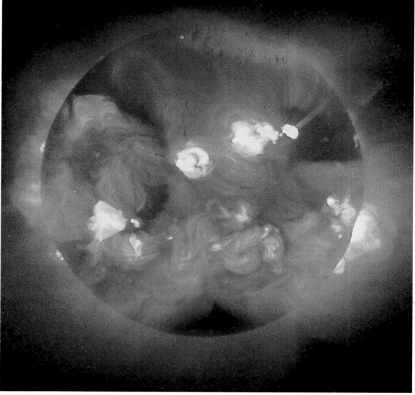

Fig. 6.9. Radio and X-ray Images of the Sun. At invisible radio (*top*) and X-ray (*bottom*) wavelengths, the Sun shows parallel bands of strong, localized emission near sunspot maximum; it is radiated by dense, hot material trapped in coronal loops in the vicinity of sunspots. The radio image was obtained (26 September 1981) at 20 centimeters wavelength using the Very Large Array located near Socorro, New Mexico; the peak brightness temperature of the radio image is two million degrees. The X-ray image was obtained a decade later (12 November 1991) with the Soft X-ray Telescope (SXT) aboard the Yohkoh satellite of the Japanese Institute of Space and Astronautical Science (ISAS). The SXT was prepared by the Lockheed Palo Alto Research Laboratory (LPARL), the National Astronomical Observatory of Japan, and the University of Tokyo with the support of NASA and ISAS. [Courtesy of George Dulk, Dale Gary and Tim Bastian, University of Colorado (*top*) and LPARL and NASA (*bottom*)]

thermore, the corona transports very little heat into the photosphere because the intensity of the corona's radiation is only about a millionth of the photospheric radiation intensity.)

So, radiation cannot resolve the heating paradox, and we must look for alternate sources of energy. Possible mechanisms involve either the kinetic energy of moving material or the magnetic energy stored in magnetic fields. Indeed, the photosphere seethes with motion even when the Sun is in a quiet, inactive state, and magnetism threads its way through the entire solar atmosphere. Unlike radiation, both of these forms of energy can flow from cold to hot regions.

For several decades, sound waves provided a widely-accepted explanation for heating the low-density, million-degree corona. In 1948–49, for example, Ludwig Biermann, Evry Schatzman and Martin Schwarzschild independently proposed that the high coronal temperature could be maintained by acoustical noise produced by the solar convection. The up and down motion of the piston-like cells, called granules, will generate a thundering sound in the overlying atmosphere, in much the same way that a throbbing high-fidelity speaker drives sound waves in the air. The sound (acoustic) waves should accelerate and strengthen as they travel outward through the increasingly rarefied solar atmosphere until supersonic shocks occur that resemble sonic booms of jet aircraft. It was thought that these shocks would dissipate their energy rapidly and perhaps generate enough heat to account for the high-temperature corona.

The majority of the sound waves are reflected back and remain trapped in the Sun (see Sect. 4.1), but a small percentage of them manage to slip through the photosphere. The internal convection currents move in and out, like a beating heart, producing a roaring sound, like the ocean surf crashing into shore. Some of the sound waves probably leak through and dissipate their energy rapidly within the chromosphere, thereby generating large amounts of heat.

The low chromosphere does indeed seem to be heated by sound waves that are generated in the convection zone and dissipated by shocks in the chromosphere, perhaps by waves with periods mainly between 2 and 4 minutes and within small magnetic channels where the chromosphere brightens. This is generally consistent with the fact that other stars with outer convection zones have chromospheres, while for stars without outer convection zones no chromosphere is detected. However, since the Skylab observations of ultraviolet spectral lines in the 1970s, it has become apparent that there is very little acoustic energy left over by the time the shock waves reach the upper chromosphere, and sound waves apparently cannot reach the corona. (The steep temperature and density gradient in the transition region would reflect sound waves, keeping them from propagating into the corona.)

Magnetic fields probably play a pivotal role in heating the solar corona. The X-ray emission of the high-temperature gas is brightest within active regions, where the magnetic field is strongest, and it is intense magnetism that molds the corona, producing its highly-structured, in-homogeneous shape. The other key ingredient for coronal heating is change. The dynamic corona is magnetically linked to, and driven by, the underlying photosphere and convection zone whose turbulent motions can push the magnetic fields around and shuffle them about. Moreover, the ever-changing, magnetized corona is continually evolving, with new magnetic loops rising up from inside the Sun and old ones decaying or dissipating away.

So, recurring themes in explaining the searing temperatures of the solar corona are highly-structured magnetism and turbulent change, but the exact way in which magnetic energy is transformed into heat is still a matter of dispute. There are several possible scenarios. Magnetic, heat-producing waves can be launched along magnetic fields by twisting or shaking their foundations. Alternatively, slow deformations and re-arrangements of the coronal magnetism can generate electrical currents that flow along the magnetic fields; these currents may heat the electri-fied and resistive gas in much the same way that current heats the fila-ment of a light bulb or an element in an electric heater. According to a third hypothesis, sudden, localized magnetic coupling, merging or re-connection, brought about by encounters of oppositely directed magne-tism, may also result in the release of stored magnetic energy as heat.

Coronal magnetic fields are constantly displaced and distorted in re-sponse to motions in the convection zone. When a magnetic field is dis-turbed, a tension acts to pull it back, generating magnetic waves that can propagate upward and dissipate energy in the corona. These waves can be transported along the same magnetic fields that sculpt the corona's features.

Radio signatures of outward-propagating magnetic waves, called Alf-vén waves after the theoretician Hannes Alfvén who first described them mathematically, have been detected within the solar corona at distances of up to ten solar radii from the Sun. However, the Alfvén waves may propa-gate right through the corona without dumping enough energy into it. (In technical terms, these waves are not easily damped and they may not dissi-pate their energy quickly enough to noticeably heat the corona.)

Although Alfvén waves may be required to explain heating that takes place in distant regions on the Sun, other magnetic interactions may be required to heat the gas within the closed magnetic structures of active regions. Their intense X-ray emission has the largest heating require-ments, and their association with the strongest magnetic fields can not be accidental. Magnetic loops emerging in active regions are shunted to and fro by turbulent convective motions below the photosphere. When the magnetic fields are sufficiently distorted, they may collapse catas-trophically to form a new configuration with thin current sheets in which energy can be rapidly dissipated.

Radio and X-ray images do provide evidence for contorted magnetic fields and electric currents in the corona. However, coronal heating may not be simply due to magnetic stresses at the coronal level, resulting from the random motion of their footpoints in the photosphere. The high-resolution images show no detectable X-ray emission above sunspot umbrae, where the magnetic field is most intense; the hot, highly-filamented magnetized loops instead descend to the surrounding penumbra. This suggests that magnetic fields by themselves do not heat the corona, and that driving motions internal to small-scale bundles of magnetism may be crucial.

The buildup of coronal magnetic stress by underlying motions may be suddenly and catastrophically released by magnetic reconnection. This may occur when emerging magnetic loops interact with pre-existing ones, or when twisted magnetism reconnects all by itself. Magnetic reconnection results in a rapid change to a more stable magnetic configuration, with the release of stored magnetic energy that can heat the corona.

According to a scenario outlined by Eugene Parker, the steady heating of the hot solar corona can be regarded as the cumulative result of a large number of intermittent, small-scale, sudden and low-level explosions, or flares, each releasing magnetic energy in small, isolated magnetic loops. The numerous, unresolved, explosive events, called nanoflares or microflares, occur at seemingly random locations, that are presumably related to chaotic motions and stressing. They are assumed to be much more numerous than large eruptions, and to combine to generate the high-temperature corona. (The greater heating rate for active regions is explained by greater energy release per nanoflare in these regions, perhaps because of the more intense magnetic fields.)

Parker's hypothesis is supported by observations of localized, transient impulsive heating and dynamic events that occur in the transition region or corona. High-resolution ultraviolet observations from space have revealed myriad, tiny, high-speed jets of matter, explosively expelled from the transition region. Horizontal motions take magnetism to the edges of the supergranulation convective cells where the explosive events occur. This is also where one finds narrow, jet-like spicules and excess chromospheric heating.

Many small, magnetized X-ray loops, called bright points, are also turning on and off within active regions (also see Sect. 6.4 – the so-called X-ray bright points are not points at all, but instead consist of relatively small magnetic loops.) The flaring loops are much more frequent and substantially less intense than major solar flares. The transient X-ray brightening of the small loops is also more frequent for hotter active regions and less frequent in cooler ones, suggesting heating by the nonsteady injection of pre-heated mass or by recurrent, low-level explosive activity.

And why don't the magnetic fields keep on heating the Sun's corona up until it explodes? The free electrons and protons make the corona a very good conductor – about twenty times better than copper. The coronal heat is therefore easily carried out into space or back down into the Sun. In fact, some of the hot coronal material expands out into colder interplanetary and interstellar space, carrying heat with it and keeping the temperature at a million degrees. However, not all of the corona flows out from the Sun; much of it is constrained within magnetized loops that can prevent expansion into space.

6.3 CLOSED CORONAL LOOPS AND OPEN CORONAL HOLES

The million-degree corona emits high-energy X-ray radiation that is totally absorbed by the Earth's dense atmosphere. However, splendid images from space, taken at X-ray wavelengths, can continuously reveal the hot, three-dimensional geometry of the corona across the full disk of the Sun. The Sun's surface is too cool (about 6000 degrees Kelvin) to emit substantial X-ray radiation, so it appears dark under the million-degree corona.

The first X-ray pictures of the Sun were obtained in 1960 by Herbert Friedman and his colleagues at the Naval Research Laboratory during a brief 5-minute rocket flight. (Sounding rockets are still used to test solar instruments and technology for subsequent use in orbiting satellites.) These crude, early images were replaced with high-resolution X-ray photographs taken and returned to Earth by astronauts from NASA's Skylab only thirteen years later, during a 9 month period in 1973–74. Most recently, the Yohkoh, or "sunbeam" spacecraft, launched on August 30, 1991 by the Japanese Institute of Space and Astronautical Science (ISAS), has provided millions of images of the Sun at invisible X-ray wavelengths that are almost as sharp and clear as pictures made in visible wavelengths from the ground.

As seen by an X-ray telescope, the solar corona is far from uniform and symmetrical. It consists of bright incandescent loops, starkly revealing magnetic fields that loop across the Sun's surface and hold the corona in place, as well as dark coronal holes, low-density regions where the solar magnetism opens out into interplanetary space. From Skylab came the first clear pictures of the ubiquitous coronal loops, that continuously alter their shape and form in response to changing magnetic fields, as well as the dark, extended open regions. The rapid, uniform sequence and wide dynamic range of Yohkoh's X-ray images, which are sent by radio link from space to Earth, have shown that the coronal material and magnetic fields change and shift their shapes on all possible spatial and temporal scales, with effects felt throughout the solar atmosphere.

So, the X-ray telescopes in space have shown that the corona is stitched together by bright, thin magnetized loops, which provide the woven fabric of the entire corona. The closed loops rise into the corona from regions of positive magnetic polarity and then turn back into a region of negative polarity in the underlying solar photosphere. By some as-yet-unknown process, material is concentrated to higher densities and temperatures within these loops, so they emit X-rays more intensely than their surroundings. This intense X-ray emission thus outlines the magnetic shape and structure of the Sun's outer atmosphere.

Instruments aboard Skylab kept the magnetized loops under close scrutiny at both ultraviolet and X-ray wavelengths, recording their appearance in different spectral lines. These are the permitted lines emitted by ionized atoms within coronal loops, and not the forbidden lines detected at visible wavelengths. Because a given stage of ionization occurs within a narrow range of temperature, the different spectral lines can be used to tune in coronal loops at particular temperatures between 100 thousand and several million degrees Kelvin, thereby dissecting their hidden structure layer by layer.

Relatively cool loops, imaged at ultraviolet wavelengths (Fig. 6.10), are located high above active regions, arching over smaller, lower-lying, hotter ones anchored in sunspots and emitting intense X-ray radiation. Relatively small, cool ultraviolet loops are also sometimes nestled between the legs of longer, hotter X-ray ones. So, the magnetic structures one sees in the corona depend very much on how you look at them, with different spectral lines revealing otherwise-invisible loops at different temperatures and heights.

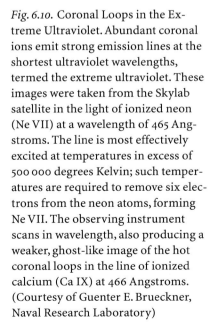

Fig. 6.10. Coronal Loops in the Extreme Ultraviolet. Abundant coronal ions emit strong emission lines at the shortest ultraviolet wavelengths, termed the extreme ultraviolet. These images were taken from the Skylab satellite in the light of ionized neon (Ne VII) at a wavelength of 465 Angstroms. The line is most effectively excited at temperatures in excess of 500 000 degrees Kelvin; such temperatures are required to remove six electrons from the neon atoms, forming Ne VII. The observing instrument scans in wavelength, also producing a weaker, ghost-like image of the hot coronal loops in the line of ionized calcium (Ca IX) at 466 Angstroms. (Courtesy of Guenter E. Brueckner, Naval Research Laboratory)

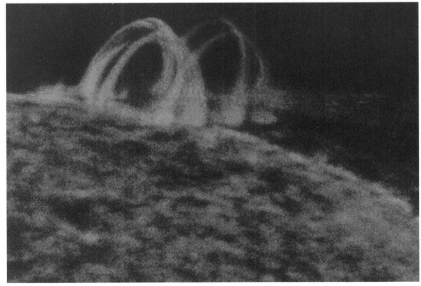

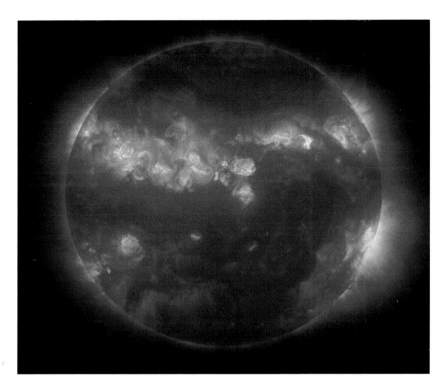

Fig. 6.11. Detailed Coronal Structure. This full-disk, soft X-ray image shows incredible details of coronal structure in and around active regions. It was taken with a Normal Incidence X-ray Telescope (NIXT) during a 5-minute rocket flight on 11 September 1989. By focusing on monochromatic radiation at the wavelength of 63.5 Angstroms, a multilayer coated X-ray mirror can be used with a higher angular resolution over a wider field of view than grazing incidence telescopes. In active regions the soft X-ray radiation at this wavelength is dominated by the emission of ionized iron (Fe XVI) formed at 3 million degrees; in the quiet corona ionized magnesium (Mg X) formed at 1 million degrees dominates the radiation. (Courtesy of Leon Golub, Smithsonian Astrophysical Observatory)

By focusing on a single X-ray line, incredibly sharp and detailed images can be obtained using an ultra-smooth mirror with multilayer coatings to increase its X-ray reflectivity. Such X-ray images, obtained during 5-minute flights of NASA sounding rockets, show structure down to 0.75 arc-seconds, or about 500 kilometers across (Fig. 6.11). They indicate that the bright, million-degree coronal gas is constrained within very thin loops, only a few arc-seconds across, with lengths that can be more than a hundred times their width, and that these magnetically active structures are very complex and highly tangled. The high-resolution images also indicate that the hot magnetized loops terminate in areas of enhanced chromospheric emission, including sunspot penumbrae, but that no hot loops connect to sunspot umbrae. The multilayer technology holds strong promise for longer observations with future instruments aboard satellites.

The thin, magnetized coronal loops confine the hottest, densest, X-ray emitting gas within active regions above sunspots. Here the magnetic pressure usually dominates the gas pressure, so the ionized gases can only move within their magnetic cage. The coronal loops therefore act like magnetic conduits, or pipes, through which the hot, dense gas flows but cannot escape.

The variety of magnetic geometries detected in X-ray images from the Yohkoh spacecraft have astonished solar physicists. Although the

Fig. 6.12. Magnetic Shapes. This X-ray image shows bright loop-like emission, presumably from hot gas trapped by strong, dipolar magnetic fields within active regions, and fainter magnetic structures with contorted, twisted geometries that may be shaped by electric currents in the corona. Relatively faint and long magnetic loops also connect active regions to distant areas on the Sun, or emerge within quiet regions away from active ones. This image was taken on 11 June 1992 with the Soft X-ray Telescope (SXT) aboard the Yohkoh mission of the Japanese Institute of Space and Astronautical Science (ISAS). The SXT was prepared by the Lockheed Palo Alto Research Laboratory (LPARL), the National Astronomical Observatory of Japan, and the University of Tokyo with the support of NASA and ISAS. (Courtesy of LPARL and NASA)

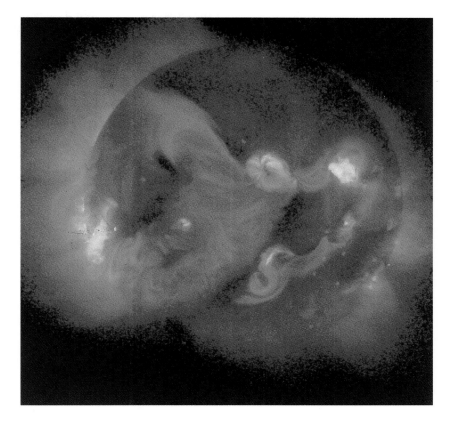

brightest loops, that are located within active regions, usually have a simple dipolar shape, less-intense, distorted, non-dipolar X-ray shapes have been found outside active regions. The irregular loops seem to be carrying electric currents along the loop, generating their own magnetism and distorting the overall magnetic shape (Fig. 6.12). Some coronal loops are contorted, others appear to explode! Large-scale magnetic conduits connect active regions, sometimes on opposite sides of the solar equator, or join active regions to more distant areas on the Sun; some of them seemingly rise up out of nowhere, providing long-distance connections where there are no active regions.

In contrast to the dense, bright areas, the corona also contains less dense regions called coronal holes. These so-called holes have so little material in them that they are not very luminous and are difficult to detect. They therefore appear as large dark areas on X-ray images seemingly devoid of X-ray radiation (Fig. 6.13). Coronal holes are nearly always present at the Sun's poles, where they appear to be predominant. They sometimes grow toward the equator like a spreading bald spot, occasionally stretching from pole to pole, and then subside poleward again. At solar minimum, coronal holes can cover a large part of the polar regions, where conspicuous polar rays or plumes are found in eclipse photographs.

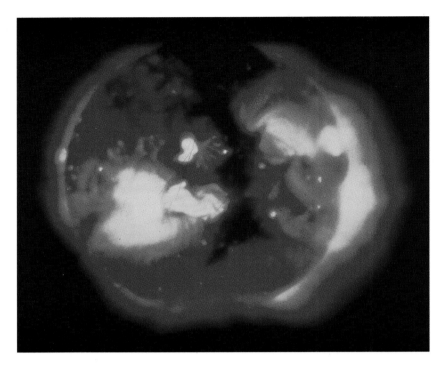

Fig. 6.13. Coronal Holes. This X-ray image shows a coronal hole as a large, dark feature that extends from the north pole (*top*) down through the middle of the solar disk. The magnetic fields above coronal holes stretch out radially and do not arch directly back to the Sun. Hot, electrified gas therefore pours outward from coronal holes, creating high-speed gusts in the solar wind. Because the gas is moving out, the gas density is lower in coronal holes and the X-ray emission is reduced. In contrast, the most intense X-ray emission comes from glowing, rope-like loops within active regions; strong magnetic fields confine the hot, dense X-ray emitting gas in these regions. Smaller regions, called bright points, are spread almost uniformly across the disk. (A Skylab image taken in April 1974, Courtesy of the Solar Physics Group, American Science and Engineering, Inc.)

Unlike the arched and closed magnetism of coronal loops, the coronal holes have relatively weak and open magnetic fields that stretch out radially into interplanetary space and do not return directly to another point on the Sun. (Of course, the fields eventually turn around, but they are distended to great distances from the Sun.) The normally constraining magnetic forces relax and open up in the coronal holes to allow an unencumbered outward flow of electrically charged particles into interplanetary space (Fig. 6.14, also see Sect. 6.5). Solar particles therefore pour out of coronal holes along the open magnetic field lines, apparently unabated, keeping their density low.

Coronal holes appear like a dark, empty void, as if there were a hole in the corona, but the rarefied coronal holes are not completely empty. Low levels of radio, ultraviolet and X-ray radiation indicate that small amounts of hot, million-degree material remain, but with a density of less than 10 percent of that found in the ubiquitous coronal loops. The relatively low density accounts for the dark, dim shapes of coronal holes in X-ray images.

Elongated coronal holes, stretching from pole to pole, sometimes maintain their vertical orientation for months. This suggests that the equatorial regions within coronal holes rotate at the same rate as the polar regions; elsewhere in the outer atmosphere the equator spins at a faster rate than the poles. This lack of slippage or shear, known as differential rotation, could be explained if the weak, large-scale magnetic fields

Fig. 6.14. Magnetic Fields Near and Far. In the low solar corona, strong magnetic fields are tied to the Sun at both ends, trapping hot, dense electrified gas within magnetized loops. Far from the Sun, the magnetic fields are too weak to constrain the outward pressure of the hot gas, and the loops break open to allow electrically-charged particles to escape, forming the solar wind and carrying magnetic fields away. (Courtesy of Newton Magazine, the Kyoikusha Company)

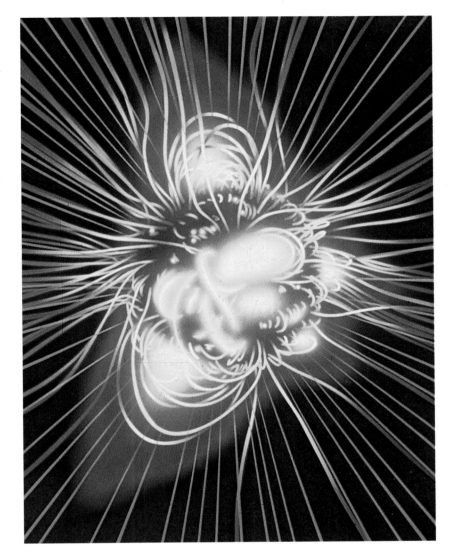

in coronal holes are linked to the deep subsurface layers below the convection zone, where the Sun might rotate like a rigid ball or solid body. Alternatively, the rotation of coronal holes could be controlled from above, instead of from below; the dark regions could mark the footpoint areas of rigidly rotating, global magnetic fields that extend high into the corona and drift relative to the underlying Sun. This might explain why some coronal holes rotate rigidly, especially at sunspot minimum, while others may participate in differential rotation.

In either event, the coronal holes are neither constant nor permanent; they appear, evolve and die away in periods ranging from a few weeks to several months. So coronal holes continuously change in shape and form, like everything else on the Sun.

6.4 THE DYNAMIC EVER-CHANGING CORONA

The apparently serene Sun, an unchanging disk of brilliant light to the casual eye, has no permanent features but is a volatile, ever-changing incandescent ball of gas. The apparently steady corona that is frozen into a single image or photograph is therefore an illusion. The dynamic, ever-changing corona is in a continued state of metamorphosis, always varying in brightness and structure on all detectable spatial and temporal scales. Like most of the rest of the Universe, the Sun's outer atmosphere is never still, and there is no such thing as a quiet, inactive corona.

One result of the Yohkoh investigation was, in fact, to demonstrate that the corona is much more active than had previously been thought. Yohkoh's rapid, sequential images show that the corona is in a constant state of agitation, always adjusting to the shifting forces of magnetism (Fig. 6.15). For instance, coronal loops within active regions (the sites of disruptive solar eruptions or flares) come and go while also calmly expanding, in some cases almost continually. Such expanding loops apparently contribute to mass loss from the Sun and other stars.

The hot, dense coronal loops within active regions can be magnetically reconfigured in hours or less, producing changes in their X-ray shape and form. Hot X-ray emitting gas is occasionally propelled out of active regions like a jet, travelling through long, well-collimated, pre-existing loops to remote locations such as the polar regions with their permanent coronal holes. And within active regions, numerous tiny X-ray loops are turning off and on every ten minutes or so, like the blinking lights of a Christmas tree.

Other small magnetic bipolar structures are found in coronal holes and elsewhere. They are called X-ray bright points because of their intense X-ray emission and relatively small size, about 10 000 to 40 000 kilometers across. X-ray bright points vary in intensity on time scales of a few minutes to hours, and they can be associated with increased X-ray emission in much larger structures that link the so-called points to remote areas on the Sun.

Indeed, the faint corona seen outside solar active regions is not an isolated, inactive entity, but is also highly dynamic, with varying brightness and continual magnetic changes. Coronal loops create magnetic conduits, sometimes as long as the Sun is wide, that can be suddenly and mysteriously filled with hot gas and just as quickly emptied. Localized changes trigger disturbances that cascade across the Sun like avalanches – as if the corona were poised on the brink of instability. The slightest magnetic disturbance can temporarily throw the entire corona out of control! It then globally adjusts to the change, relaxing to a less-agitated, lower-energy state, but is never quiet.

Fig. 6.15. The Ever-Changing Corona. ▷ This sequence of X-ray images illustrates rapid restructuring of the solar corona. All four images include the Sun's southeast limb curving down from upper left to lower right. The first image (*upper left*) was taken about 3 hours before the second one (*upper right*), followed by a third image 1.5 hours later (*bottom left*) and a final one 10 minutes after that (*bottom right*). The trigger for these changes was the eruption of a filament. These images were taken on 26 February 1992 with the Soft X-ray Telescope (SXT) aboard the Yohkoh mission of the Japanese Institute of Space and Astronautical Science (ISAS). The SXT was prepared by the Lockheed Palo Alto Research Laboratory (LPARL), the National Astronomical Observatory of Japan, and the University of Tokyo with support from NASA and ISAS. (Courtesy of LPARL and NASA)

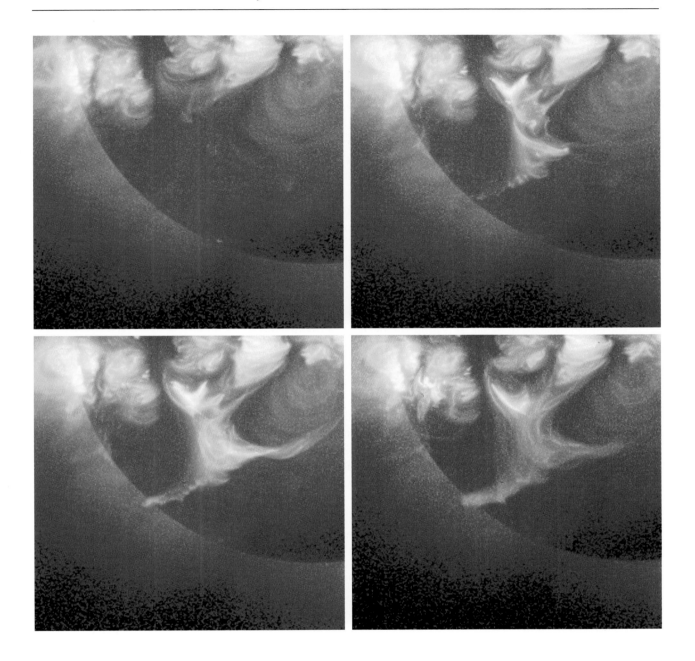

6.5 THE ETERNAL SOLAR WIND

The Sun is continuously blowing itself away! It boils off the outer layer of its atmosphere, spewing it into space and filling the solar system with a perpetual flow of electrified matter called the solar wind. That is, the Sun's corona is expanding into interplanetary space, forming a ceaseless wind. Unlike any wind on Earth, the solar wind is an electrified stream of charged particles.

Every second the Sun blows away about a million tons of mass that must be replaced from below, but this is a small amount compared with the enormous total mass of the Sun. At the present rate, it would take ten billion years for the Sun to lose only 0.01 percent of its mass by the solar wind, and the Sun will evolve into a giant star long before it blows away completely.

So the space between the planets is not completely empty; it is filled with pieces of the Sun. The eternal solar wind, a rarefied plasma or mixture of protons and electrons, streams out radially in all directions from the Sun, and the planets move through this wind as if they were ships at sea. The solar gale brushes past the planets and engulfs them, carrying the Sun's corona out to interstellar space. The radial, supersonic outflow thereby creates a huge bubble of plasma, with the Sun at the center and the planets inside, called the heliosphere, from *helios* the Greek word for the Sun.

The existence of an eternal solar wind flowing in all directions from the Sun was inferred decades ago from the observed motions of comet tails. The comets appear unexpectedly almost anywhere in the sky, moving in every possible direction, but with tails that always point away from the Sun. Moreover, the ions in comet tails always stream away from the Sun with velocities many times higher than could be caused by the weak pressure of sunlight (Fig. 6.16).

In the early 1950s, the German astrophysicist Ludwig Biermann proposed that streams of electrically charged particles, called corpuscular radiation, poured out of the Sun at all times and in all directions to shape the comet tails. (In previous decades, it was realized that the so-called corpuscular radiation from the Sun causes geomagnetic storms and polar auroras.)

Summing up his work in 1957, Biermann concluded that:

> The acceleration of the ion tails of comets has been recognized as being due to the interaction between the corpuscular radiation of the Sun and the tail plasma. The observations of comets indicate that there is practically always a sufficient intensity of solar corpuscular radiation to produce an acceleration of the tail ions of at least about twenty times solar gravity ... . There is no reason why similar interactions should not be expected between the solar corpuscular radiation and a stationary interplanetary plasma; it follows that an assumed interplanetary cloud would not remain stationary.[27]

Thus, the ion tails of comets act like an interplanetary wind sock, demonstrating the existence of a continuous, space-filling flow of charged particles from the Sun. As a result, comet tails always point away from the Sun, so they travel head first when approaching the Sun and tail first when departing from it. (To be exact, the electrified solar wind deflects charged ions in the comet tails, always sending them away from the Sun,

Fig. 6.16. Comet Tails. This photograph shows the curved dust tail and straight ion tail of Comet Myros; both comet tails point away from the Sun. The pressure of the Sun's light, called radiation pressure, gives the dust particles an outward push, creating a broad arc that resembles a scimitar. In contrast, a wind of charged particles, the solar wind, accelerates the ions to high velocities and pushes them into the relatively straight ion tails. The comet travels head first when it approaches the Sun and tail first when moving away. (Courtesy of Lick Observatory)

but the radiation pressure of the sunlight suffices to blow away the un-ionized cometary dust, creating the shorter curved dust tail – also see Fig. 6.16).

We also know that the million-degree corona is so hot that it cannot stand still. Indeed, the solar wind consists of an overflow corona, which is too hot to be entirely constrained by the Sun's inward gravitational pull. The hot gas creates an outward pressure that tends to oppose the inward pull of the Sun's gravity; and at great distances, where the solar gravity weakens, the hot protons and electrons overcome the Sun's gravity and accelerate away to supersonic speed, like water overflowing a filled bathtub or a dam.

So, the solar corona is really the visible, inner base of the solar wind, and the solar wind is just the hot corona expanding into cold, vacuous interstellar space. In 1957, geophysicist Sydney Chapman demonstrated mathematically that the Sun's million-degree corona must extend beyond the Earth's orbit, even if it is gravitationally held to the Sun. In the following year, Eugene Parker, a young astrophysicist at the University of Chicago, theorized that the solar wind inferred from Biermann's account of comet observations is no more than the expanding corona.

Parker showed from hydrodynamics that the corona must expand rapidly outward because it is extremely hot, and that as the outer corona disperses, it will be replenished by gases welling up from below. The hydrodynamics shows that the expansion would begin slowly near the Sun,

FOCUS 6B

SOHO

The primary goal of the Solar and Heliospheric Observatory, or SOHO, mission is to provide answers to two basic, unsolved questions of current solar physics. What heats the Sun's corona to its million-degree temperatures, and what accelerates the Sun's eternal wind, or expanding corona, to supersonic velocities of hundreds of kilometers per second? These goals will be achieved both by remote sensing of the invisible solar atmosphere with high resolution spectrometers and telescopes, and by *in situ* measurement of the composition and energy of the resulting solar wind and the energetic particles that propagate through it.

The other goals of SOHO include a study of the Sun's interior structure and dynamics that result in energy flow and magnetic fields at the base of the corona. These goals will be accomplished by helioseismological methods and the measurement of solar irradiance variations (also see Sect. 4.5).

The SOHO mission is a joint venture of the European Space Agency (ESA) and NASA. The spacecraft, which is expected to be launched in 1995, will obtain an uninterrupted view of the Sun from a point known as the L1 Lagrangian point where the gravitational forces of the Earth and the Sun balance one another. The L1 point is located approximately 1 percent of the distance from the Earth to the Sun, or at about 1 500 000 kilometers from the Earth.

SOHO is part of a larger effort known as the International Solar-Terrestrial Physics (ISTP) science initiative. ISTP aims, through coordinated exploration of the space regions neighboring Earth (geospace), to understand the behavior of the solar-terrestrial system, and, therefore, the way in which the Earth's environment responds to the varying solar output. The solar-terrestrial interaction is fully discussed in the last three chapters of this book.

where the solar gravity is the strongest, and then continuously accelerate outward into space, gaining speed with distance and reaching the supersonic velocities needed to account for the acceleration of comet tails. This would create a strong, persistent, solar wind, forever blowing at speeds of hundreds of kilometers per second out through the solar system and beyond.

Any doubts about the existence of the solar wind were removed by *in situ* (Latin for "in original place", or literally "in the same place") measurements made by instruments on board NASA's Mariner II spacecraft on its way to Venus in 1962. The solar wind has subsequently been directly sampled for more than three decades, over three cycles of solar activity, detecting a continuous solar wind that is much hotter, thinner and faster than any wind on Earth. The million-degree, low-density, electrified gas flows past the Earth with densities of about 10 particles per cubic centimeter at steady speeds ranging from 300 to 700 kilometers per second and with gusts that can move twice as fast. Moreover, spacecraft travelling well beyond the farthest planet continue to detect the solar wind. We also know today that the solar wind's density, temperature and velocity vary dramatically over periods of just a few hours as well as over the decade duration of the solar activity cycle. These changes reflect the ever-changing state of activity on similar time scales in the Sun's corona.

However, the exact sources of this gusty wind and the forces that propel it with such tremendous energy remain largely unknown. We do not understand the basic driving mechanism of the solar wind; and it is not known why the solar wind is sometimes accelerated to around 700 kilometers per second (the tenuous, high-speed solar wind). (Curiously the fast wind has a lower density, and the slow wind has a higher density, so the total energy needed to overcome gravity and to supply kinetic energy to the wind is the same for both the slow-speed and the high-speed winds.)

Of course, the wind is accelerated to about 300 kilometers per second because there is a hot, million-degree corona, and the corona exists because something heats it; but we still do not know exactly what heats the corona or what gives the solar wind an extra boost to 700 kilometers per second. The heating of the solar corona presents one of the most fundamental, unsolved problems of contemporary solar physics (also see Focus 6B).

We may not know exactly how the solar wind is accelerated, but we do know where some of it originates. Gases welling up from beneath the corona apparently flow out along the weak and open magnetic fields within coronal holes, creating high-speed streams in the solar wind. In these large, X-ray-dark holes, the Sun's magnetism stretches radially outward with little divergence, producing a fast lane for the electrified wind. The high-speed solar wind squirts out of the nozzle-like coronal holes, like water out of a fire hose, and the larger the hole, the higher the speed of the wind.

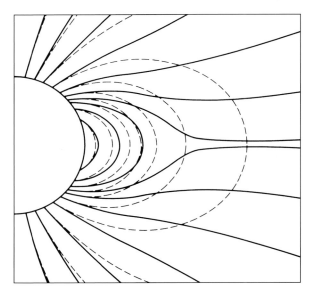

Fig. 6.17. Magnetic Threads. Theoretical cross section of the magnetic field lines expected in the Sun's corona during a minimum in solar activity. It represents a simple dipole (*dashed lines*) whose magnetic field lines have been pulled outward by the solar wind into a million-degree corona. The high-speed wind escapes along the open magnetic field lines from the polar regions (*top* and *bottom*). Here, the transition from closed to open field lines at the equator occurs at two and a half times the Sun's radius where the domination of the solar wind becomes most apparent; the slow-speed wind might originate in these equatorial regions (*right center*). [Adapted from Gerald W. Pneuman and Roger A. Kopp, Solar Physics *18*, 258–270 (1971)]

Equatorial coronal holes can expel high-speed particles directly toward the Earth; but the fast solar wind also gushes out of the polar regions and eventually crosses the Earth's orbit (Fig. 6.17). The polar winds spread out to fill all the space around the Sun with their flow. Since large coronal holes are almost always present at each pole, they might supply most of the high-speed solar wind.

But what is the source of the slow-speed wind? The basic mystery is the heat source of the corona. Given the observed temperatures, the ordinary slow-speed solar wind is a consequence of hydrodynamics, and the more the magnetic field fans out the faster the wind. The second mystery is the high-speed stream with velocities of 600 to 800 kilometers per second, or about twice the expected velocity of the slow-speed wind.

The distant reaches of the expanding corona can be probed using remote radio sources that fluctuate, or scintillate, in much the same way that stars twinkle when seen through the Earth's atmosphere. The radio waves are perturbed when they pass through the solar wind, producing a hazy, blinking and distorted image. It's something like looking at a light from the bottom of a swimming pool.

The radio data have shown that the solar wind does not blow evenly or steadily. It varies over time, exhibiting squalls and calms, suggesting that transient activity feeds the slow-speed component of the solar wind. The radio scintillations also indicate that the average wind velocity increases from the solar equator to the higher latitudes toward the solar poles where coronal holes are most often observed.

Out in the corona, a dynamic tension is set up between the magnetic field and the charged particles. The field tends to push the particles around, but the particles produce currents that tend to resist and move

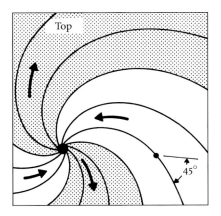 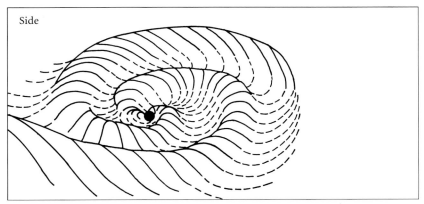

Fig. 6.18. Warped Spirals. If the interplanetary magnetic field could be viewed from the Sun's polar regions (*left*), it would have a spiral pattern that is divided into sectors of opposite magnetic polarity, like the vanes of a pinwheel. At the Earth's location (*small black dot*), the pulling effect of the solar wind is about equal to the twisting effect of the Sun's rotation, so the stretched-out field makes an angle of about 45 degrees with the radial direction from the Sun. In a three-dimensional view (*right*), the sector structure is represented by a wave-like current sheet that divides magnetic fields directed away from the Sun and those directed toward it. The current sheet rotates with the Sun and sweeps regions of opposite magnetic polarity past the Earth.

the field about. In the low corona within active regions, strong magnetism wins and the hot particles are confined in coronal loops. But weaker magnetic fields are found in coronal holes and in the outer corona, at large distances from the Sun, both above active regions and at the base of helmet streamers, where the magnetic field decreases in strength. The abundant low-energy charged particles of the solar wind drag these weaker portions of the Sun's magnetism along with them as they stream far beyond the most distant planets.

The expansion of the solar wind combined with the Sun's rotation determines the magnetic structure of interplanetary space, and thereby establishes the magnetic pathways for energetic particles leaving the Sun. While one end of the solar magnetic field remains firmly rooted in the solar photosphere, the other end is extended and stretched out into space by the solar wind. As the wind streams radially outward, the Sun's rotation bends the radial pattern into a spiral shape within the plane of the Sun's equator, coiling the magnetism up like a tightly-wound spring (Fig. 6.18).

The spiral magnetic field creates an interplanetary highway that can connect the site of a solar eruption, or flare, to the Earth. Energetic charged particles, that are fewer in total number and much greater in energy than those in the solar wind, are hurled out from the Sun during these brief eruptions, creating powerful gusts in the solar wind. If they occur in just the right place, near the west limb and the solar equator, the energized material will connect to the interplanetary Archimedian spiral and travel along it to the Earth in about half an hour, threatening astronauts or satellites. The spiral magnetic pattern has, in fact, been confirmed by tracking the radio emission of charged particles thrown out during such solar flares, as well as by spacecraft that have sampled the interplanetary magnetism near the Earth.

Interplanetary space probes have also measured the direction, or polarity, of the field, showing that it is divided into sectors of alternating magnetic polarity. The field is directed toward or away from the Sun in adjacent sectors, like oppositely aligned bar magnets. For every field line leaving the Sun, another threads its way back through an adjacent sector of opposite magnetic polarity. So there is no such thing as completely open magnetic fields on the Sun. They are just greatly distended into magnetic sectors.

Large interplanetary regions of opposite magnetic polarity are separated by a magnetically neutral layer, or current sheet, along which current can flow freely. This neutral current sheet lies approximately within the plane of the Sun's equator, but the sheet is not flat. Its base meanders across the photosphere like the curved seam of a baseball. When the Sun rotates, the current sheet wobbles up and down like a warped record, sweeping regions of opposite magnetic polarity past the Earth (also see Fig. 6.18). At the time of reduced sunspot activity, when the helmet streamers are stretched out along the solar equator, there appears to be little warping of the current sheet, but during periods of high solar activity the sheet can be severely warped.

6.6 ULYSSES

Until recently, spacecraft have only been able to directly measure a limited, two-dimensional section, or slice, of the solar wind within the plane in which the Earth orbits the Sun. This plane, called the ecliptic, is tilted only 7 degrees from the plane of the solar equator, so the Earth and its artificial satellites never manage to get more than 7 degrees of solar latitude above or below the solar equator in the course of a year. Interplanetary spacecraft also usually travel within the ecliptic, both because they normally rendezvous with another planet and also because their launch vehicles obtain a natural boost by travelling in the same direction as the Earth's spin and in the plane of the Earth's orbit. In fact, there are currently no launch vehicles capable of generating the thrust needed to get a spacecraft out of the ecliptic plane.

Yet, the Ulysses spacecraft, launched by NASA's Space Shuttle Discovery in October 1990, has begun to assess the total solar environment for the first time. It has travelled outside the ecliptic plane, permitting us to escape the confines of its two-dimensional view and directly sample the entire three-dimensional realm of the solar wind. Indeed, the Ulysses mission, a collaboration between NASA and the European Space Agency (ESA), is exploring the Sun's wind over the full range of solar latitudes, including previously unexplored regions above the Sun's poles.

The spacecraft is named after Ulysses, Prince of Ithaca and the greatest explorer of Greek mythology; his exploits are recounted in Homer's *Iliad*, and his long wandering before he reached home, in the *Odyssey*. In Dante Alighieri's *Inferno* Ulysses recalls his restless desire to explore the unknown world, exhorting his friends

> To venture the unchartered distances;
>
> to feel life and the new experience
>
> Of the uninhabited world behind the Sun . . . .
> To follow after knowledge and excellence.[28]

Like its namesake, the Ulysses mission is on a voyage of exploration to a new domain of space.

At the time of launch, Ulysses was the fastest man-made object in the Universe, with a velocity of 15.25 kilometers per second, but it was then moving at only about a third of the velocity required to propel the spacecraft up and out of the ecliptic. To bend the flight path, Ulysses was sent out to the planet Jupiter, whose immense gravity was used to deflect the spacecraft in slingshot fashion out of the ecliptic into a highly inclined orbit back inward toward the Sun's south pole. (At Jupiter's distance, the Sun's gravitational pull is less than at Earth, so the velocity needed to fling Ulysses poleward is substantially less, by a factor of 2.5.) So, in February 1992, fourteen months after launch, Ulysses completed the planned rendezvous with Jupiter, and was thrust back and out on a trajectory that carried it first over the Sun's southern pole, in mid-1994, then up and over the northern pole a year later.

During all that time, Ulysses continuously measured the solar wind streaming from the Sun, like a remote space-weather station. It showed how the velocity and magnetic field of the solar wind change on two fronts – distance from the Sun and distance from the ecliptic plane. Ulysses' measurements of the solar wind velocity as a function of latitude, or distance from the ecliptic, at sunspot minimum (Fig. 6.19) seems

Fig. 6.19. Ulysses. Solar wind speeds measured by the Ulysses spacecraft. In mid-1992, when the spacecraft was still within the ecliptic, the slow-speed component of the solar wind was dominant, with a velocity of about 400 kilometers per second. As Ulysses moved out of the ecliptic toward the solar poles, the Sun's rotation alternately swept the low-latitude, slow wind and the higher-latitude, high-speed component past the spacecraft, producing a modulation between the two velocities with a period of 25.5 days. By mid-1994, Ulysses had arrived above the south pole of the Sun where the solar wind retains only the highest velocity component of about 760 kilometers per second. Thus, during the Ulysses measurements, the slow wind originates at low solar latitudes, between plus and minus 30 degrees, while the fast wind is expelled from well-developed polar coronal holes. At the time labeled CME the fast wind was augmented by a powerful gust associated with a coronal mass ejection. (Courtesy of John L. Phillips, Los Alamos National Laboratory)

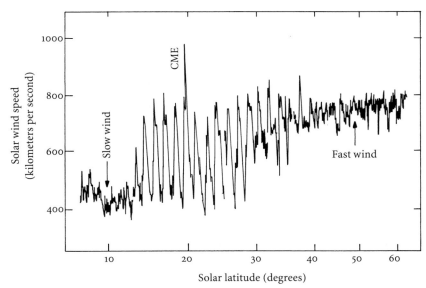

to confirm that the high-velocity wind originates mainly in pronounced coronal holes at the solar poles and spills over into the equatorial regions, and that the slower component of the solar wind is produced at or near the plane of the Sun's equator, at least during the time of measurement when active regions and coronal streamers may have been concentrated at low solar latitudes. However, the exact origin of the slow-speed component of the solar wind is still a matter of controversy.

The global distribution of the Sun's magnetic field will become more disordered at a later phase in the solar cycle, when activity is at a maximum. Ulysses may then discover that there is less distinction between the slow and fast components of the solar wind, when coronal holes are less pronounced and explosive solar activity may help drive the fast wind. But already, even before all the results are in, we would agree with Joachim du Bellay (1522–1560) that:

Happy he who like Ulysses has made a glorious voyage.[29]

6.7 THE DISTANT FRONTIER

All of the planets are immersed in the solar wind that becomes increasingly rarefied as it spreads out into space. It moves past the planets and beyond the most distant comets to mark the very edge of the solar system. Thus, the entire solar system is bathed in the hot gale that blows from the Sun, creating the heliosphere, a vast region centered on the Sun and enclosed by the interstellar medium.

Within the heliosphere, physical conditions are dominated, established, maintained, modified and governed by the Sun.

The Sun's wind thins out as it expands; by the time it has reached the Earth's orbit it is nearly a perfect vacuum by terrestrial standards. As it moves outward and spreads into a greater volume, the density of the solar wind decreases even further and eventually blends with the gas between the stars. (Its density decreases as the inverse square of distance from the Sun.)

What marks the outer boundary of the heliosphere, where the solar wind ebbs and the cold of interstellar space begins? The solar wind fills an ever-larger volume of space until it is no longer dense or powerful enough to repel the ionized matter and magnetic fields coursing between the stars. It is this turbulent boundary, called the heliopause, that marks the edge of our solar system.

Instruments on the twin Voyager spacecraft, cruising far beyond the outermost planets in distant unchartered territory, have recorded signals from the solar wind's termination in space. Strong shock waves associated with intense eruptions on the Sun have plowed into the cold interstellar gas at the heliopause, generating a hiss of radio noise detected by the

Fig. 6.20. Edge of the Solar System. ▷
Remote Voyager spacecraft, now headed on an escape trajectory from the Sun, will never return; they have picked up signals thought to come from the heliopause, where the solar system ends and interstellar space begins. At the heliopause, the pressure of the solar wind balances that of the interstellar medium. Although the heliopause marks the outer edge of the heliosphere, the region of the Sun's control, it is not expected to be spherical. The interstellar medium probably flows past the Sun with enough speed to generate a bow shock on the upstream side, and to pull the heliosphere into a long magnetotail in the direction of flow. When a dense, high-speed gust of solar particles, and their associated shock wave, hit the heliopause, they generated an intense radio hiss detected by the widely separated spacecraft, enabling them to estimate that the edge of our solar system is somewhere between 116 and 177 times farther away from the Sun than is Earth. (Ian Worpole, © 1993 Discover Magazine)

1 Interstellar wind
2 Source of radio emissions
3 Voyager 1
4 Voyager 2
5 Heliopause
6 Solar wind
7 Solar system

remote spacecraft (see Fig. 6.20). So powerful and pervasive were the radio signals that they were equally strong when detected almost simultaneously by the two spacecraft, even though they were travelling in different directions and separated by 44 AU. (One astronomical unit, or
1 AU, is the mean distance between the Earth and the Sun.)

Thirteen months before the spacecraft detected the radio hiss, unusually intense eruptions on the Sun generated some of the largest interplanetary disturbances ever observed. From the measured speed of the
resulting disturbance, and the thirteen months it took to travel to the heliopause and generate the radio signals, the heliopause has been located
somewhere between 116 and 177 AU.

The two Voyager spacecraft keep going and going. Launched in 1977,
they were at 53 AU (Voyager 1) and 41 AU (Voyager 2) from the Sun in
1993. The hardy vehicles are expected to chart the wind's full extent,
sending faint radio messages back to Earth through the year 2015 when
Voyager 1 and 2 will be at 130 AU and 110 AU, respectively.

Closer to home, space physicists are concerned about the impact of
powerful solar eruptions on the Earth's environment in space.

Cosmic Synchromy 1913–14. In this painting, Morgan Russell develops the theme found in Robert Delaunay's portrayal of the Sun and Moon (see frontispiece to Chap. 3), extending arcs of pure color into space. They seem to resonate throughout the Universe like myriads of cosmic rainbows. (Courtesy of Munson-Williams-Proctor Institute, Museum of Art, Utica, New York. Oil on canvas, 16 $\frac{1}{4}$ × 13 $\frac{1}{8}$ in)

CHAPTER 7

The Violent Sun

7.1 ENERGETIC SOLAR ACTIVITY

Without warning, the relatively calm solar atmosphere can be torn asunder by sudden outbursts of a scale unknown on Earth. Catastrophic events of incredible energy can produce transient brightenings, called solar flares, that flood the solar system with intense radiation from X-rays to radio waves. The powerful flares are easily observed at these invisible wavelengths, where they briefly dominate the Sun's output and sometimes outshine all other astronomical sources (Fig. 7.1).

Solar flares are brief, catastrophic outbursts of awesome power and violence. (The word flare means a sudden, rapid and intense variation in brightness.) In minutes, the disturbance spreads along concentrated magnetic fields, releasing stored magnetic energy equivalent to billions of nuclear explosions and raising the temperature of Earth-sized regions to tens of millions of degrees. Solar flares are therefore hotter than the corona. Sometimes they temporarily go out of control and lose equilibrium, becoming hotter than the core of the Sun itself for a short period of time.

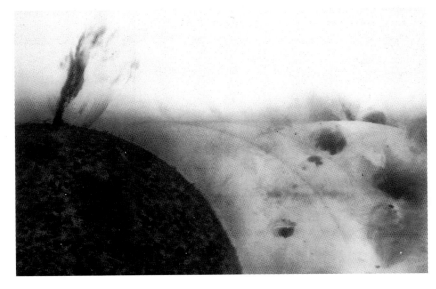

Fig. 7.1. Ultraviolet Eruption. Solar eruptions and coronal loops are easier to detect at invisible wavelengths where the photosphere does not dominate the radiation. An explosion is caught in the ultraviolet light of ionized helium (He II at 304 Angstroms) that forms at about 80 000 degrees Kelvin (*left side*), while numerous million-degree coronal loops appear (*right side*) in the invisible ultraviolet light of 14 times ionized iron (Fe XV at 284 Angstroms). In these images the Sun has been tipped on its side; the coronal loops and the eruption actually occur in equatorial active regions. (Images from the Skylab satellite, courtesy of Guenter E. Brueckner, Naval Research Laboratory)

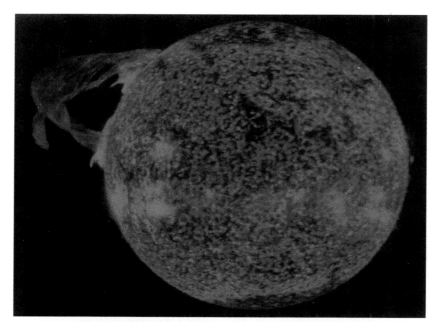

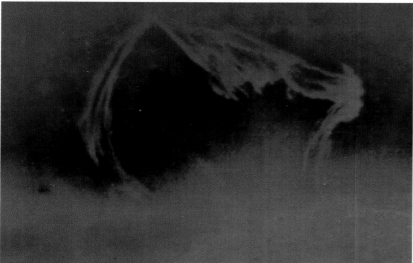

Fig. 7.2. Pieces of the Sun. Magnetic loops and energetic material are hurled far out into interplanetary space. On the scale of these pictures, the Earth would be only slightly larger than the period at the end of this sentence. The twisting, eruptive prominences arch about 400 000 kilometers above the visible solar disk. These images were taken from the Skylab satellite in the invisible ultraviolet light of singly ionized helium (He II at 304 Angstroms); this spectral line forms at a temperature of about 80 000 degrees Kelvin. (Courtesy of Guenter E. Brueckner, Naval Research Laboratory)

In another type of energetic solar activity, prominences or filaments, elongated structures that stretch up to halfway across the visible solar surface, suddenly and unpredictably open up and expel their contents, defying the Sun's enormous gravity (Fig. 7.2). The erupting filaments are associated with coronal mass ejections, giant magnetic bubbles that expand as they propagate outward from the Sun to rapidly rival it in size (Fig. 7.3). These violent eruptions throw billions of tons of material into interplanetary space. Their associated shocks accelerate and propel vast quantities of high-speed particles ahead of them.

Fig. 7.3. Coronal Mass Ejection. Billion-ton bubbles of hot gas grow larger than the Sun in just a few hours. This time sequence of coronagraph images, covering just over 4 hours, illustrates some of the principal features of many coronal mass ejections – the presence of a bright, outer loop of material, followed by a dark cavity, under which is visible a bright loop-like structure identified with erupting prominence material. These images were taken with the coronagraph aboard the Solar Maximum Mission (SMM) spacecraft. (Courtesy of Arthur J. Hundhausen, High Altitude Observatory and NASA)

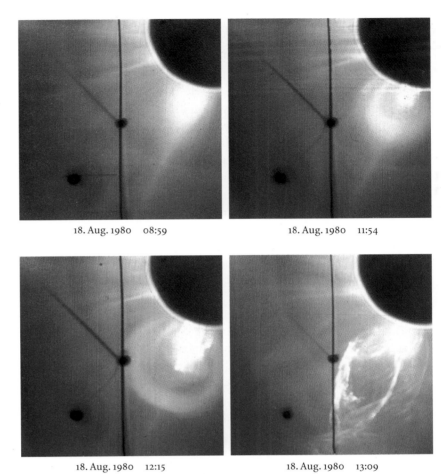

18. Aug. 1980 08:59

18. Aug. 1980 11:54

18. Aug. 1980 12:15

18. Aug. 1980 13:09

The rates of occurrence of solar flares, erupting prominences and coronal mass ejections all vary with the 11-year cycle of magnetic activity (see Sect. 5.3), becoming greatest at sunspot maximum. Truly outstanding flares are infrequent, occurring only a few times a year even at times of maximum solar activity; like rare vintages, they are denoted by their date. Flares of lesser magnitude occur much more frequently; several tens of such events may be observed on a busy day near the peak of the 11-year solar cycle of magnetic activity. One or two coronal mass ejections balloon out of the corona per day, on average, during this activity maximum, and the rate decreases by about an order of magnitude by sunspot minimum.

All kinds of solar activity therefore seem to be related to the sudden release of stored magnetic energy, but the exact relation between them is unclear. Large coronal mass ejections are sometimes followed by solar flares, for example, but not all flares are caused by mass ejections. Flares occur much more frequently.

Solar eruptions can seriously disrupt the Earth's environment. Intense radiation from powerful solar flares can travel to the Earth in just 8 minutes, altering its outer atmosphere, disrupting long-distance radio communications, and affecting satellite orbits. Very energetic particles, accelerated during the flare process or by the shock waves of coronal mass ejections, arrive at the Earth within an hour or less (for energies above 10 MeV); they can endanger unprotected astronauts or destroy satellite electronics. (One MeV is a million electron volts, which is about a million times the energy associated with visible light.) The coronal mass ejections arrive at the Earth as a dense cloud of magnetic fields, electrons and protons one to four days after a major eruption on the Sun, resulting in strong geomagnetic storms with accompanying aurorae and electrical power blackouts. All of these effects are of such vital importance that national centers employ space weather forecasters and continuously monitor the Sun from ground and space to warn of threatening solar activity. (These solar-terrestrial effects are discussed in greater detail in Sects. 8.7, 8.8 and 8.9.)

Energetic solar activity has even broader implications, for the entire Universe is full of similar cataclysms. The dark night sky, with its steady stellar beacons, may give the impression of stillness and serenity, but it is actually a place of cosmic violence. Indeed, the acceleration of energetic particles, with the associated emission of powerful, invisible radiation, characterizes much of the Universe, from black holes and pulsars to the most distant quasar. Investigations of the Sun's energetic activity provide important insights to understanding violence throughout the Universe, for only the Sun is close enough and bright enough to be studied with sufficient detail.

7.2 SOLAR FLARES

Our perceptions of solar flares have evolved with the development of new methods of looking at them. Despite the powerful cataclysm, most solar flares are not, for example, detected on the bland white-light face of the Sun; they are only minor perturbations in the total amount of emitted sunlight. So, the visible output of the Sun remains remarkably constant, uniform, stable and apparently dependable in spite of the activity. For this reason flares were relatively late to be discovered, and a complete study of them had to await investigations at invisible radio and X-ray wavelengths.

The first report of a solar flare was made independently, by two English observers, Richard C. Carrington and Richard Hodgson, on 1 September 1859. Carrington was observing the forms and positions of sunspots when he suddenly saw two brief, intense sources of light above a complex sunspot group. According to his account:

> Two patches of intensely bright and white light broke out … .
> I hastily ran to call some one to witness the exhibition with me, and
> on returning within 60 seconds, was mortified to find that it was
> already much changed and enfeebled. Very shortly afterwards the
> last trace was gone.[30]

Carrington's friend, Mr. Hodgson, chanced to be observing the Sun at the same time and confirmed this first account of a white-light flare in the astronomical literature. Carrington noted that this event took place in the middle of a very intense magnetic storm at the Earth (it lasted from 28 August to 4 September and was detected at the Kew Observatory in London), but a definite link between individual solar events and terrestrial disturbances was not established until the following century.

Carrington noticed that the sunspots were precisely the same after the sudden conflagration as they were before, leading him to conclude that:

> The phenomenon took place at an elevation considerably above the
> general [visible] surface of the Sun, and, accordingly, altogether
> above and over the great [sunspot] group in which it was seen
> projected.[31]

To this day, more than 135 years later, no solar observer has yet recorded a definitive change in the magnetic fields at the photospheric level during a solar flare, and it is generally believed that they originate in the overlying low coronal atmosphere.

Carrington observed a relatively rare event, in which a flare's light was enhanced sufficiently over the background photospheric sunlight to be visible by contrast. Most solar flares are not so conspicuous in the combined colors, or white light, of the Sun.

A new perspective, that demonstrated the frequent occurrence of moderate solar flares, was made possible when they were observed in the monochromatic, hydrogen light (at 6563 Angstroms) of the chromosphere. Routine observations of solar flares at this wavelength did not begin until after George Ellery Hale's invention of the spectrohelioscope in the 1920s. This instrument allowed the Sun to be imaged in spectral lines such as hydrogen-alpha in which solar flares can be easily observed visually. Hale's spectrohelioscope led in the 1930s to regular patrol observations of the Sun, nowadays using automatic cameras and telescopes. For more than half a century, astronomers throughout the world have carried out this vigilant, continued flare patrol, like hunters waiting for the sudden flash of game birds.

These systematic patrol observations have provided a two-dimensional picture of solar flares. They detect one slice through the Sun's atmosphere at the chromospheric level, showing that this part of a solar flare can consist in simplest form of either bright point-like kernels or two extended, parallel flare ribbons. Since the hydrogen-alpha observations consisted largely of morphological descriptions of size, shape and location, they have been dubbed solar dermatology by the critical.

A fundamental understanding of the physical processes responsible for solar flares had to wait until they were detected at invisible wavelengths from both the ground and space, beginning in the 1960s and 1970s and continuing with increased sophistication to the present day. When combined, the visible and invisible observations provide a full three-dimensional view of solar flares.

Although flares usually produce only minor perturbations in the visible white light of the Sun, the flaring radio and X-ray emission is frequently several thousand times more energetic than the Sun's normal radiation at these wavelengths. Relatively small telescopes can therefore be used to detect invisible flares. The radio and X-ray radiation indicate the presence of high-speed particles and heated gas arising from the rapid release of energy during solar flares.

Nonthermal electrons accelerated during solar flares travel within the magnetic conduits of active regions, interacting with the magnetic fields and producing intense radio emission. These invisible signatures of solar flares are sometimes called radio bursts to emphasize their brief, energetic and eruptive characteristics. (The electrons have been dubbed nonthermal because their speeds can be close to the velocity of light, and such a pronounced acceleration cannot occur when a gas is heated by thermal processes.)

The magnetic fields bend the paths of the nonthermal electrons into a circle, resulting in the emission of a narrow beam of synchrotron radiation that spins around like the beam of a lighthouse (Fig. 7.4). It is called

Fig. 7.4. Synchrotron Radiation. An electron does not cross a magnetic field line, but instead spirals around it. If the electron moves perpendicular to the field, the magnetic force acts like a rubber band, pulling the particle back and constraining it to move in small circles about the magnetic field line. The electron can, however, move freely in the direction of the magnetic field. This combination of the free motion along, and constrained circular rotation around, the magnetic field shapes the electron's trajectory into a helix that spirals around a magnetic field line. When the electron moves rapidly at velocities near that of light, it sends out electromagnetic waves called synchrotron radiation after the synchrotron particle accelerator where it was first observed. Unlike thermal radiation, the nonthermal synchrotron radiation is most intense at long, invisible radio wavelengths rather than short visible ones. Because the electron moves along the magnetic field line, its radiation is emitted preferentially in one direction, so it is linearly polarized.

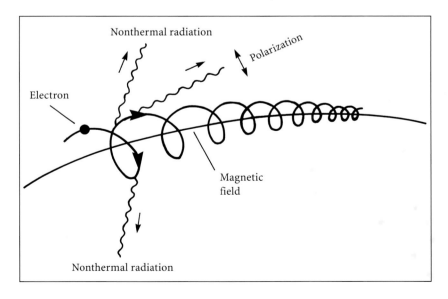

Nonthermal radiation

Polarization

Electron

Magnetic field

Nonthermal radiation

Fig. 7.5. Yohkoh Catches a Flare. A Soft X-ray Telescope (SXT) aboard the Yohkoh mission captured this image a few hours after powerful flares erupted from an active region (AR 6891 on 25 October 1991). The postflare X-ray loops (*lower left*) are much brighter than coronal loops found in quiescent, or non-flaring, active regions on other parts of the Sun. This X-ray image has a wide dynamic range of about 130 000, obtained as a composite of three images with exposures of 10 milliseconds, 80 milliseconds and 2.6 seconds, thereby enabling the detection of both faint and bright coronal loops. Dark coronal holes are also present at both poles (*top* and *bottom*). Yohkoh is a mission of the Japanese Institute of Space and Astronautical Science (ISAS); the SXT was prepared by the Lockheed Palo Alto Research Laboratory (LPARL), the National Astronomical Observatory of Japan, and the University of Tokyo with the support of NASA and ISAS. (Courtesy of Keith T. Strong, LPARL)

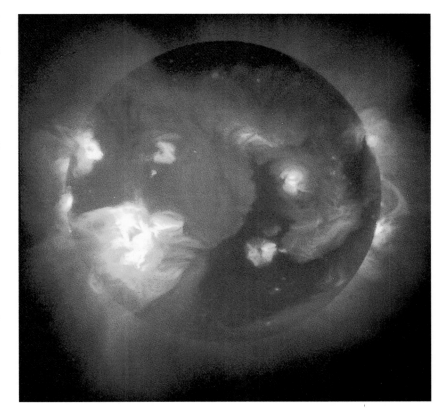

synchrotron radiation after the man-made, ring-shaped synchrotron particle accelerator where it was first observed. (The name "synchrotron" refers to the synchronous mechanism that keeps the particles in step with the acceleration as they circulate in the ring.)

Because solar flares have very high temperatures, the bulk of their radiation is emitted at X-ray and ultraviolet wavelengths. This radiation is absorbed in the Earth's atmosphere, so astronomers observe it with telescopes in outer space. During a large flare, soft X-ray emission briefly dominates the background radiation of even the brightest coronal loops (Fig. 7.5). X-rays with relatively long wavelengths and low energies are called soft X-rays to distinguish them from the very energetic, short-wavelength hard ones.

The soft X-ray radiation is named *bremsstrahlung*, the German word for braking radiation. It is produced when free electrons pass near ions (also see Sect. 6.2). Their electrical interaction bends the path of the electrons and speeds them up; and the electrons radiate soft X-rays in the process. The bremsstrahlung is also called thermal radiation since it depends on the random thermal motion of the hot electrons, in contrast to the more energetic electrons emitting nonthermal synchrotron radiation at radio wavelengths.

Satellites that are specifically intended to investigate flare emission
have now been launched to coincide with three successive maxima in
the 11-year cycle of solar activity (see Focus 7A). Skylab soft X-ray photographs taken in 1973–74 showed that the low solar corona is shaped
and constrained by the ubiquitous coronal loops (see Sect. 6.3), and indicated that solar flares can occur in coronal loops anchored in underlying sunspots. Such soft X-ray images reveal the presence of a very hot
gas with temperatures of up to 25 million degrees, and also outline or
trace the magnetic configuration of the coronal material. The loop-like
soft X-ray flare emission either links the hydrogen-alpha flare kernels
or forms arcades of emission bridging the hydrogen-alpha flare ribbons
in the underlying chromosphere.

FOCUS 7A

Observing Solar Flares From Space

Our knowledge of the basic physics of solar flares has increased dramatically as the result of observations from space. They began with primitive
instruments aboard sounding rockets and continued with increasingly
sophisticated telescopes, to the present day with Earth-orbiting satellites
specifically designed to study flares. NASA's Skylab mission, for example,
obtained high-resolution images of solar flares at ultraviolet and X-ray
wavelengths in 1973–74; it was followed by the Solar Maximum Mission,
or SMM, launched by NASA in 1980 and the Japanese Hinotori spacecraft, launched in 1981; both detected the high-energy, invisible radiation
of solar flares.

The SMM instruments included spectrometers at soft and hard X-
ray wavelengths, designed to image the bremsstrahlung of energetic
flares as well as temperature-sensitive and density-sensitive spectral
lines. Hinotori, meaning *fire-bird* in Japanese, included a better hard X-
ray telescope that had the temporal and spatial resolution required to
position and resolve the flaring hard X-ray emission. (SMM was repaired
by space-shuttle astronauts in 1984 and continued to obtain data until
1989; a malfunctioning data recorder terminated Hinotori observations
in 1982.)

Then in 1991, Japan launched the Yohkoh, or *sunbeam*, spacecraft. It
carries a coordinated set of four co-aligned instruments designed to
study transient high-energy phenomena in solar flares at soft X-ray, hard
X-ray and gamma-ray wavelengths.

Interplanetary spacecraft also now probe the solar wind and detect
energetic particles thrown out into space in association with solar flares
and coronal mass ejections.

Fig. 7.6. Flares in the Corona. Solar flares often occur at the tops of magnetic loops that connect underlying sunspots. Electrons are rapidly accelerated there during the early stages of the flare, emitting radio radiation delineated by the white contours. These 10-second snapshot maps were obtained with the Very Large Array (VLA) at a wavelength of 20 centimeters. The high-energy electrons are rapidly channeled down the magnetic loops, colliding with the chromospheric hydrogen atoms that emit the hydrogen-alpha light shown in the underlying photographs. [Courtesy of Kenneth R. Lang, Tufts University (VLA) and the Big Bear Solar Observatory, California Institute of Technology (hydrogen-alpha)]

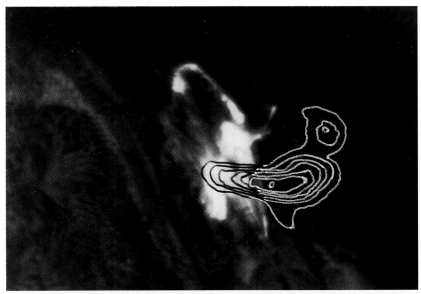

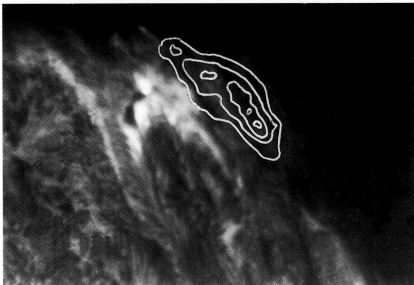

Giant arrays of radio telescopes can zoom in at the very moment of a solar flare, showing that they are often ignited in compact, small-scale structures at the tops of coronal loops (Fig. 7.6). In just a few seconds, electrons can be accelerated within the magnetized loops to nearly the velocity of light, emitting powerful radio signals. Thus, nonthermal particle acceleration seems to be localized at single or multiple loop tops, but no one knows exactly what is happening up there.

We are nevertheless certain that the short-lived solar flares unleash their vast power from a relatively small volume within active regions, the magnetized atmosphere in, around and above sunspots. The largest solar flares cover but a few tenths of a percent of the solar disk, and are intimately related to the magnetized coronal loops that link sunspots of opposite magnetic polarity.

Hard X-ray images, also taken from space, record high-speed electrons slamming down into the dense lower atmosphere, showing where the electrons are losing energy by collisions. Nuclear reactions can also occur when subatomic flare particles, such as protons, collide with the atmosphere. The nuclear reactions are identified by their gamma-ray lines with energies in the MeV range. Flare particles even produce antimatter, in the form of positrons, that annihilate with their material counterpart, the electrons, producing gamma-ray signatures (at 0.511 MeV) that have been observed from space.

The three-dimensional perspective of flaring coronal loops, derived from multiple-wavelength observations at hydrogen-alpha, radio and soft and hard X-ray wavelengths, has led to a canonical model of solar flares involving particle acceleration and energy release at the apex of coronal loops. This model is described in Sect. 7.4.

7.3 FLARE RADIATION FROM ENERGETIC ELECTRONS

What have we learned from this modern space-age investigation of solar flares? Within one second or less, electrons and protons are accelerated to speeds approaching the velocity of light, and then beamed down into the Sun or out into space. As the nonthermal electrons spiral down along magnetic channels, they generate intense radio emission. The downward-moving particle beams strike the lower, denser chromosphere, like a bullet hitting a concrete wall, with the electrons producing bremsstrahlung at hard X-ray and gamma-ray wavelengths and the protons and heavier ions initiating nuclear reactions like a colossal atom smasher. The hard X-rays often originate from the feet of flaring magnetized loops, nearly co-spatial with the hydrogen-alpha emission. Following this impulsive deposition of energy into the denser regions of the solar atmosphere, material from these regions explodes upwards and rebounds.

Soft X-rays provide a different perspective, describing the hot, thermal gas prior to, during and after the solar flare, and establishing its temperature, density and magnetic configuration. So, what you see depends on how you look at it, and solar flares are no exception.

We have also learned that invisible eruptions on the Sun can be explained as a three-part phenomenon that lasts a few minutes or tens of minutes (Fig. 7.7). All three stages of a solar flare have a characteristic radio signature, and high-resolution radio data indicate that they sometimes occur in nearby, but spatially separate, coronal loops.

Fig. 7.7. Phases of a Solar Flare. A solar flare is recorded at radio wavelengths with the Very Large Array, or VLA. Its time profile consists of the precursor, impulsive and decay, or post-impulsive, phases. Although energetic hard X-ray radiation is also emitted during the impulsive phase, the less-energetic soft X-ray emission often builds up slowly and becomes most intense during the decay phase. VLA snapshot maps made every few seconds indicate that the three phases in this case originated in spatially-separate, but nearby, coronal loops. (Courtesy of Robert F. Willson and Kenneth R. Lang, Tufts University)

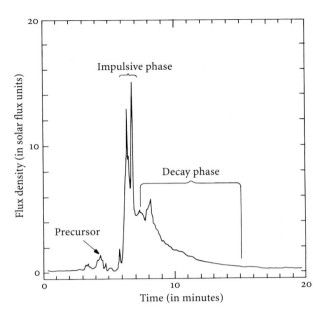

The first precursor stage triggers the release of magnetic energy that has been built up and stored in coronal loops. Evidence for this mysterious triggering mechanism is found within localized, low-level changes in the soft X-ray and radio emission of active regions; it precedes the next more-energetic impulsive phase by at most a few minutes.

At the impulsive stage, electrons and protons are accelerated rapidly and almost simultaneously to energies that exceed one MeV, and stored magnetic energy is released quickly on time-scales of seconds or less. Some of these energetic particles have been detected by spacecraft near the Earth. Signatures of this impulsive phase include intense, invisible radio bursts and hard X-ray and gamma ray emission.

During the last decay, or post-impulsive, phase, energy is being released more gradually on longer time-scales of tens of minutes. Soft X-rays slowly build up in strength and usually reach peak intensity during the post-impulsive decay phase. (The slow, smooth rise of the soft X-ray radiation resembles the integral of the time-profile of the impulsive hard X-ray emission, an effect named after Werner M. Neupert, who first noticed it in 1968.)

The high-speed electrons accelerated during solar flares can be confined within closed magnetic structures. The trapped electrons then modulate the radio emission, creating periodic variations as they move back and forth in these coronal loops (Fig. 7.8).

The electrons frequently break free of their magnetic cage and are ejected from the low corona outward along open magnetic field lines through the outer corona and interplanetary space, travelling about one third of the speed of light or at roughly 100 000 kilometers per second.

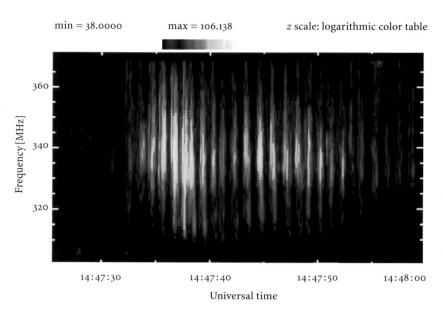

min = 38.0000   max = 106.138   z scale: logarithmic color table

Frequency [MHz]

360

340

320

14:47:30   14:47:40   14:47:50   14:48:00

Universal time

Fig. 7.8. Quasi-Periodic Pulsations. This spectrogram of a solar flare was recorded by the Ikarus radio spectrometer near Zurich, Switzerland. Time increases to the right, and the frequency increases upward. The radio emission is caused by electrons trapped in coronal magnetic fields. The flare eruption sets up low-frequency waves that propagate into the trap and modulate it. The radio pulsations are interpreted as the result of such periodic changes of the source conditions. (Courtesy of Arnold O. Benz, ETH, Zurich)

These rapid particles sometimes travel unimpeded all the way to the Earth where they are detected by instruments aboard spacecraft.

As the flare-accelerated electrons move through the coronal gas, they set it radiating at its natural vibration frequency. The local coronal electrons are displaced with respect to the more massive ions, but electrical attraction pulls the electrons back in the opposite direction. They therefore quiver and oscillate back and forth, emitting radio waves at a wavelength or frequency that depends on the local electron concentration. Ground-based radio telescopes can identify these radio signals simply by changing the wavelength or frequency to which the radio telescope is tuned.

When flare electrons move outward through the progressively more rarefied layers of the solar atmosphere, they excite radiation at longer and longer wavelengths or lower and lower frequencies. By tuning in at several wavelengths or frequencies, radio astronomers can therefore follow the outward path of electron beams, determining their speed while also measuring the progressive decrease in the electron density of the corona (Fig. 7.9).

Solar radio astronomers can additionally monitor beams of electrons that are sent down into the Sun from their birthplace in coronal loops. When successively shorter wavelengths, or higher frequencies, are tuned in, we can watch electron beams excite radiation in the denser regions below the loop tops (also see Fig. 7.9). When combined, radio signals from both the outward-moving and inward-travelling electron beams can pinpoint the coronal regions where the nonthermal electrons originate. Thus, it is believed that the flare energy is stored within coronal magnetic

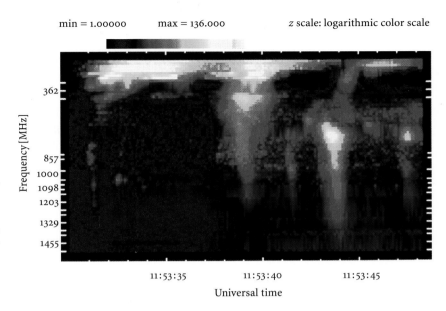

Fig. 7.9. Coronal Electron Beams. The Phoenix radio spectrometer near Zurich, Switzerland captured this record of the radio emission caused by beams of energetic electrons that propagate along magnetic field lines. Here time increases to the right, and frequency increases downward. Outward-moving beams in the solar corona emit radio waves at decreasing frequency. Their radiation drifts from lower left to upper right in the picture. Electron beams going down toward the Sun drift the other way. Both kinds of beams are visible, apparently originating at the same acceleration site. (Courtesy of Arnold O. Benz, ETH, Zurich)

fields and released from them. However, the exact method by which this stored magnetic energy is suddenly released (in just a few minutes or, in some cases, seconds), and subsequently transported throughout the solar atmosphere and beyond is far from fully understood.

7.4 ENERGIZING SOLAR FLARES

What energizes catastrophic eruptions on the Sun? They release more energy than that present in the heat of the entire corona, so they cannot draw their energy from the surrounding gas. At the time of eruption, the energy has to be present locally within the corona, and most astronomers believe it is stored in the strong magnetic fields there. Magnetic energy is slowly built up, stored in the low solar corona for long periods of time, and then abruptly released, like the sudden flash and crack of a lightening bolt from a storm cloud.

Solar flares usually occur within active regions where sunspots and coronal loops congregate, and the more magnetically complex the sunspot group, the higher the frequency of flare occurrence. Because the solar surface is constantly in motion, and because its rotation is not uniform in latitude or with depth, its embedded magnetic fields are constantly changing their shape and location. The magnetism becomes sheared, stretched, tangled and twisted by churning motions deep within the Sun, slowly accumulating stored energy in the process, in much the same way as continually twisting a rubber band increases the tension and stores energy in its kinks and bends.

The energy keeps growing until the magnetized coronal loops are pushed beyond their limits, or an outside force intervenes and destabilizes them. Then in a matter of seconds the stored energy is rapidly and violently unleashed, like the quick snap of a rubber band that has been twisted too tightly. A solar flare erupts accompanied by copious acceleration of charged particles and the emission of intense invisible radiation.

The sudden loss of equilibrium in the coronal magnetism has been likened to an avalanche in a sandpile. As sand is added, the average slope of the sandpile increases until it reaches a state where it remains approximately constant. Once this critical state is reached, addition of more sand causes avalanches of varying size which readjust the local slope, keeping the system in the same critical state. In this analogy, the stored magnetic energy slowly builds up until it is on the brink of instability, in a critical condition where further perturbations result in avalanche-like disruptions.

During the prelude to a flare, stressed magnetism causes substantial electric currents to flow along the coronal loops within an active region. The free, current-carrying (non-potential) magnetic energy is usually regarded as excess energy compared with that of the corresponding current-free (potential) magnetism. Sophisticated magnetographs can be used to specify when the photospheric magnetic field has been distorted into a current-carrying configuration, indicating that there are electric currents flowing into and out of the corona through the photosphere. It is believed that a flare results from the sudden dissipation of these components of the coronal magnetism.

A solar flare might be caused by magnetic merging, a process in which magnetic fields pointing in opposite directions come into contact, releasing magnetic energy in the form of heating and particle acceleration. (The technical name for this coupling is magnetic reconnection.) Because the Sun's convective, differentially rotating gas is in constant motion, the embedded magnetic fields are frequently brought together in the overlying corona.

The various multiple-wavelength perspectives of solar flares have been combined in a canonical model for their energy release (Fig. 7.10). Since flares apparently originate in the corona, and the ubiquitous coronal loops dominate its structure, it is perhaps not surprising that this model involves a single coronal loop. Some unknown, catastrophic instability triggers the primary energy release of a flare at the loop top, where electrons and protons are accelerated; these energetic particles produce a variety of observable emissions from different parts of the loop. The high-speed electrons emit synchrotron radio radiation at or near the loop apex during the impulsive phase of a solar flare. Energy is transported downward by the electrons and protons which follow the coronal loop's arching magnetism into the lower, denser reaches of the solar atmosphere. Here the particle beams are slowed by collisions, emitting hard X-rays and gamma rays at small localized areas that mark the loop footpoints. The rarely-seen, white-light flares are presumed to be produced by the downward impact of the nonthermal electrons.

Since the chromosphere is heated very rapidly by the accelerated particles, it cannot get rid of the excess energy and it explodes up into the corona. This may be followed by the more gradual release of energy when the coronal loop relaxes into a more stable configuration during the decay phase of a solar flare, accompanied by increased soft X-ray radiation as the flaring loop is filled with the rising evaporated material.

Although this orthodox interpretation of solar flares provides a simple framework for interpreting a wide variety of results, we are also now discovering the great diversity in solar flares. In some cases, hard X-ray radiation precedes impulsive radio emission by a few seconds, but with similar time profiles indicating a common origin for both. This suggests that high-energy electrons can propagate from the chromosphere into the corona in the reverse order from the canonical model. In addition, much more complicated magnetic geometries are usually involved in all but the simplest flares, including multi-loop structures, long arcades of loops, and interacting coronal loops. The flaring loop, or loops, may not be confined to a single active region, for impulsive energy can be trans-

Fig. 7.10. Canonical Model of a Solar Flare. A solar flare may take place within a single, isolated coronal loop whose basic structure remains unchanged throughout the flare. The impulsive flare begins with a primary energy release at the top of the coronal loop, where electrons and protons are rapidly and efficiently accelerated to speeds approaching the velocity of light. Energized particle beams transport energy downward as they stream along the curved magnetic channel, producing different radiation signatures at various parts of the loop. The high-energy electrons emit radio radiation at or near the loop top, and they produce hard X-rays and gamma rays when they penetrate deep into the chromosphere at the loop footpoints. When the particle beams enter the dense, lower atmosphere, they also produce nuclear reactions that result in gamma-ray spectral lines and energetic neutrons. This is followed by the explosive rise of heated material, called chromospheric evaporation, into the loop, accompanied by a slow, gradual increase in soft X-ray radiation.

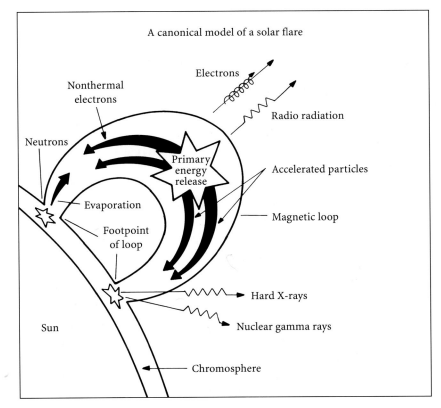

ferred to distant regions through large-scale magnetic conduits, even
across the solar equator or into the polar coronal regions. In fact, recent
work on energetic solar activity focuses on large-scale magnetic instabil-
ities that result in erupting prominences and coronal mass ejections that
are discussed in the next sections.

7.5 ERUPTING PROMINENCES

Large-scale magnetic fields hold up and insulate giant, elongated
structures filled with material that is hundreds of times cooler and
denser than the surrounding corona. They are called prominences when
seen in hydrogen-alpha photographs taken at the apparent edge of the
Sun, perhaps because they prominently stand out against the dark background, shining like fluorescent viaducts or bridges (also see Sect. 5.2).
They appear as dark, snaking features, called filaments, when projected
against the bright chromospheric disk.

So, prominence and filament are essentially two words that describe
different perspectives of the same thing. Their long sinuous forms trace

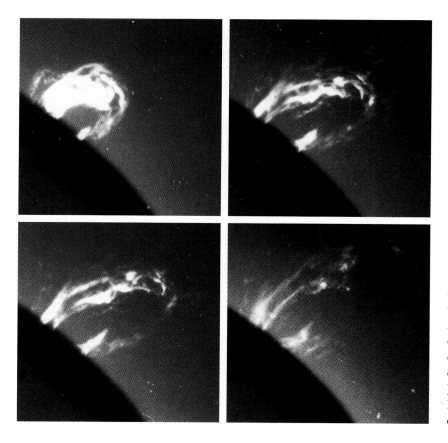

Fig. 7.11. A Prominence Erupts. Solar explosions are caught by rapid, sequential hydrogen-alpha photography. The erupting prominence had not been detected as a filament during previous days; it suddenly rose from an active region and expanded at an apparent velocity of 375 kilometers per second, or about a million miles per hour, hurling material far away from the Sun. Here the magnetic loops rise to a maximum visible extent of 360 000 kilometers in just 16 minutes. This sequence of hydrogen-alpha images was taken on the west edge, or limb, of the Sun using the automatic flare patrol heliograph at the Meudon Station of the Observatoire de Paris; the solar disk has been occulted to give a better view of the event. (Courtesy of Madame Marie-Josephe Martres, and observer Michel Bernot, Observatoire de Paris, Meudon, DASOP)

out a region of magnetic neutrality that separates large areas of opposite magnetic direction, or polarity, in the underlying photosphere. (By way of comparison, coronal loops that are anchored within individual active regions are about ten times smaller than the total length of the largest prominences or filaments.)

Prominences and filaments can hang or float almost motionless above the photosphere for weeks or months, but there comes a time when they cannot bear the strain. Then, without warning, the supporting magnetism becomes unhinged and a surprising thing happens! Instead of falling down under gravity, the stately, self-contained structures erupt, often rising as though propelled outward through the corona by a loaded spring (Figs. 7.11 and 7.12). It's as if the lid had been taken off the caged material. Cool gas is then flung outward in slingshot fashion, tearing apart the overlying corona and ejecting large quantities of matter into space.

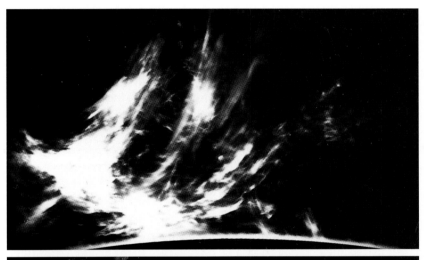

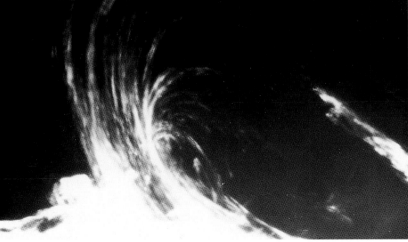

Fig. 7.12. Filamentary Detail. These coronagraph pictures, taken in the light of the hydrogen-alpha transition, show fine magnetic filaments in the legs of an erupting magnetic arch. Their orientation and structure change as the prominence erupts, perhaps as the result of the magnetic reconnection of oppositely-directed magnetic field lines. (Courtesy of Bogdan Rompolt, Astronomical Institute, Wroclaw, Poland)

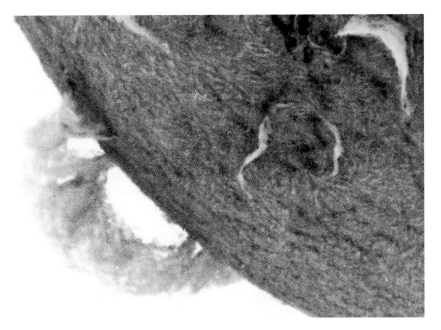

Fig. 7.13. Disparition Brusque. A prominence shines brightly at the south-east edge of the Sun in the red light of hydrogen-alpha, printed here as a negative image for contrast. It had been observed as a dark filament for weeks at a time during several previous rotations of the Sun. Then, in less than 40 hours after this picture was taken, the prominence was no longer visible. It probably rose and disappeared high in the corona within just a few hours. The French astronomers use the term *disparition brusque* to describe this sudden disappearing act of a large prominence rooted in a quiet region of the Sun. (Courtesy of Madame Marie-Josephe Martres, Observatoire de Paris, Meudon, DASOP)

These eruptive prominences, as they are called when viewed at the Sun's apparent edge, are hurled outward at speeds of several hundreds of kilometers per second, releasing a mass equivalent to that of a small mountain in just a few hours. Such an event is sometimes called a *disparition brusque* for its sudden disappearance (Fig. 7.13). Eruptive prominences can be larger, longer-lasting and more massive than solar flares.

Prominences, or filaments, often re-form in the same shape and place after an explosive convulsion. It is as if some minor irritation builds up beyond the limit of tolerance, and the magnetic structure tosses off the pent-up frustration, like a dog shaking off the rain. Long, dark filaments rise and disappear, replaced by an elongated arcade of bright X-ray loops across the initial filament position (Fig. 7.14), so their magnetic backbone regroups as before beneath the erupting prominence.

Closed magnetic loops apparently support the long filament, like parallel hammocks, at heights that are about ten times the diameter of the Earth. This arcade of closed loops is anchored in the Sun, but is opened up at the top by the rising filament, like opening the top of a jack-in-the-box. The magnetism subsequently reconnects and closes up again beneath the erupting filament or prominence, forming a new arcade of closed loops that shines in X-rays and resembles a giant's rib cage.

For a long time, erupting prominences were treated as a sort of side show in solar physics. Small dedicated groups continued to observe them, but nearly everyone else in the field ignored them. In contrast, solar flares are more easily observed at a variety of visible and invisible wavelengths, so much more observational and theoretical work has gone

Fig. 7.14. Eruptive Prominence and Arcade Formation. This is a superposition of three time-spaced radio images, taken with the Nobeyama Radioheliograph, and a negative disk image obtained with the Soft X-ray Telescope on the Yohkoh satellite at a slightly later time. The radio images show an eruptive prominence, or disparition brusque, above the disk at the north-east (*top left*) during a 1.5 hour period. The soft X-ray image shows an arcade of loops formed by magnetic reconnection after the eruption. (Courtesy of Shinzo Enome, Nobeyama Radio Observatory)

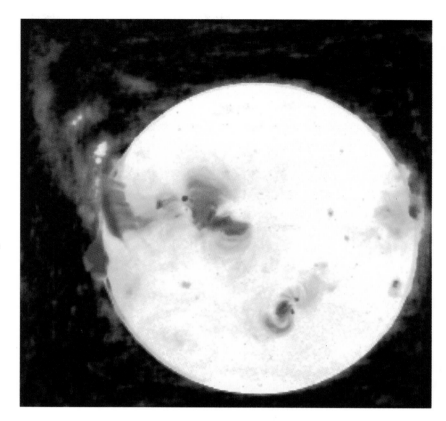

into understanding them. Then it was recently realized that disappearing prominences are closely associated with another form of energetic solar activity, the coronal mass ejections, that play an important role in solar-terrestrial interactions.

7.6 CORONAL MASS EJECTIONS

The most spectacular solar eruptions are gigantic magnetic bubbles, called coronal mass ejections, that expand outward from the Sun. Detectable only from space-borne coronagraphs – instruments that eclipse the Sun's bright disk – they were not discovered until the early 1970s. The mass ejections are seen at the apparent edge of the Sun as enormous loop-like bubbles in light scattered by coronal electrons; bright regions contain excess coronal mass while comparatively dark regions have less mass. In spite of their relatively late discovery, thousands of mass ejections have been observed routinely during the last few decades with white-light coronagraphs aboard the Orbiting Solar Observatory (OSO-7, 1972), Skylab (ATM, 1973–1974), P 78 (Solwind, 1979–1985) and the Solar Maximum Mission (SMM C/P, 1980 and 1984–1989).

The magnetic loops become unstable, carrying out billions of tons of coronal material as they lift off into space. The outward-moving coronal mass ejections stretch the magnetic field until it snaps, leaving behind only bright rays rooted in the Sun. They can expand to become larger than the Sun itself, streaming outward past the planets and dwarfing everything in their path.

Such events work only in one direction, always moving away from the Sun into interplanetary space and never falling back in the reverse direction. They often exhibit a three part structure – a bright outer loop, followed by a depleted region, or cavity, that rises above an erupted prominence (see previous Fig. 7.3). The leading bright loop, or coronal mass ejection, may be formed by a rapidly expanding, bubble-like shell that opens up and lifts off like a huge umbrella in the solar wind, piling the corona up and shoving it out like a snowplow.

Mass ejections erupt from the Sun as self-contained structures of hot material and magnetic fields. They apparently result from a rapid, large-scale restructuring of magnetic fields in the low corona. The surface distribution of coronal mass ejections over the solar activity cycle is similar to that of the other large structures on the Sun, the coronal streamers and their underlying filaments. Near activity minimum the coronal mass ejections are largely confined to equatorial regions, but near maximum they can be observed at all latitudes.

When the global magnetic fields become unstable, the coronal mass ejections erupt, carrying significant amounts of energy and matter into large volumes of interplanetary space. Five billion tons, or five million billion grams, of solar material are thrown outward during an average ejection with typical speeds of a few hundred kilometers per second. Some of them are ejected so forcefully that they move with speeds of up to 2000 kilometers per second. The energy of this mass motion is comparable to the net radiated energy of a large solar flare (about a hundred million, trillion trillion ergs).

Although subatomic particles are accelerated to high energies during solar flares, it is now thought that coronal mass ejections may be the main source of most of the energetic particles arriving at the Earth. In the currently preferred picture, strong mass ejections serve as pistons to drive huge shock waves ahead of them. The mass ejections plow into the slower-moving solar wind, like a car out of control, producing shock waves millions of kilometers across. Electrified particles energized by the shock then travel outward with it and along the broad array of magnetic field lines that intersect the shock, somewhat like surfers riding the ocean waves.

Coronal mass ejections can energize subatomic particles and open up magnetic field lines on a grand scale relatively high in the corona, eventually covering large regions of interplanetary space. In contrast, the compact, low-lying flares either confine energetic particles in closed magnetic loops or expel them within relatively narrow openings; the particles accelerated during flares subsequently follow well-defined paths described by the interplanetary magnetic spiral (see Sect. 6.5).

When coronal mass ejections were discovered in the early 1970s, it was thought that they were an explosive consequence of the bright flares; at the time flares were the most energetic eruptive phenomena known on the Sun. Then a key piece of evidence was provided by the heretofore largely ignored prominences. When astronomers began to make association studies of coronal mass ejections and solar surface activity, they found to their surprise that it was the sleeping giant prominences that, upon waking and erupting, were best associated with the mass ejections. Since erupting prominences are not usually accompanied by intense flares it was concluded that these flares were not required to drive mass ejections.

In addition, the physical size of the mass ejections dwarfs that of flares and even the active regions in which flares occur. Moreover, the departure or "lift-off" times of coronal mass ejections precede those of associated flares. Thus, the cause and effect relationship between flares and coronal mass ejections was turned on its head.

Nevertheless, flares are much more common than mass ejections. Most flares therefore probably originate in a somewhat different environment than the mass ejections, but perhaps by a similar magnetic process. The main differences between the two may just be a matter of physical size, with the compact flares occurring more often than the larger mass ejections. A comprehensive model for these closely coupled phenomena is presented in the next section.

7.7 A COMPOSITE MODEL FOR ENERGETIC SOLAR ACTIVITY

According to one scenario, large-scale, oppositely-directed magnetic fields come together and merge to energize solar eruptions. Such a magnetic configuration is suggested by the bulb-shaped base and elongated stem of a helmet streamer; they respectively consist of low-lying closed magnetic fields and magnetism that opens up to the interplanetary medium (Fig. 7.15).

A theoretical model that invokes large-scale magnetic connection has been dubbed the CSHKP model after the first letters of the last names of various researchers who have developed it over the years. (Such a model has been invoked by Hugh Carmichael in 1964, Peter Sturrock in 1968, T. Hirayama in 1974 and then Roger A. Kopp and Gerald W. Pneuman in 1976.)

Some flares detected in Yohkoh's soft X-ray images exhibit a helmet-shaped geometry when detected at the Sun's apparent edge or limb. In this X-ray structure, opposing magnetic fields stretch out and are brought together at the top of a coronal loop (Fig. 7.16). This lends sup-

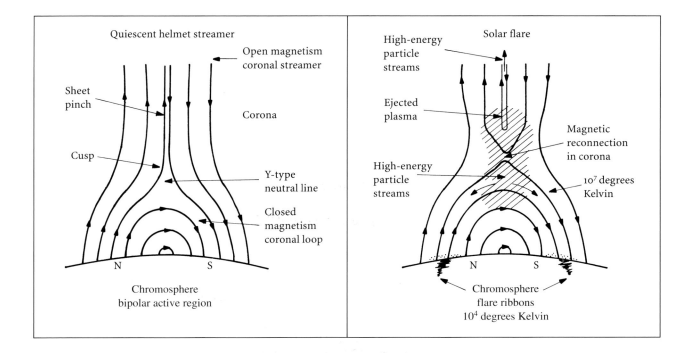

Quiescent helmet streamer

Open magnetism
coronal streamer

Sheet
pinch

Corona

Cusp

Y-type
neutral line

Closed
magnetism
coronal loop

N   S

Chromosphere
bipolar active region

High-energy
particle
streams

Solar flare

Ejected
plasma

Magnetic
reconnection
in corona

High-energy
particle
streams

10^7 degrees
Kelvin

N   S

Chromosphere
flare ribbons
10^4 degrees Kelvin

port to the CSHKP flare model in which the magnetic structures open and close like a sea anemone. The initially closed coronal loops open up – allowing the catastrophic release of energy and material stored within the coronal loops – and then close again as the magnetic fields come back together. The closed magnetic loops become buoyed up and inflated, and a coronal mass ejection results.

The mass ejection, with its accompanying erupting prominence, blows open the previously closed magnetic structure, like a hot-air balloon that breaks its tether. The magnetic cage is torn asunder, releasing pent-up magnetic energy. The associated flare is the result of the energy released by magnetic coupling of the open field lines as they pinch below the rising prominence (Fig. 7.17). Nonthermal electrons accelerated at the site of magnetic connection produce the radio and hard X-ray radiation of a solar flare. Flare loops are subsequently detected at soft X-ray and hydrogen-alpha wavelengths, shining from the newly closed magnetic loops in the flare's thermal afterglow.

A loss of equilibrium in the large-scale (global) magnetic field configuration apparently drives a coronal mass ejection outward. The corona may not be altogether surprised by this development, and may be expecting it. Many ejections occur in pre-existing coronal streamers that bulge and brighten for one to several days before erupting. The helmet streamer is then blown away by the ejection and disappears. So, it looks as if coronal mass ejections could be magnetically controlled and driven as the theoretical model suggests.

Fig. 7.15. Magnetic Reconnection in the Corona. The cross-sectional geometry of a helmet streamer (*left*) suggests a method in which oppositely directed magnetic fields can come together, annihilate each other, and reconnect (*right*). According to this version of the CSHKP model (see text), proposed by Peter Sturrock in 1968, the magnetic reconnection results in high-energy particle acceleration, ejected plasma and the formation of two bright ribbons in the chromosphere.

Fig. 7.16. X-ray Helmet Structure. A large helmet-type structure is seen in the south-west (*lower right*) quadrant of this soft X-ray image. It formed after a coronal mass ejection apparently blew open a closed magnetic configuration that subsequently reconnected to form the soft X-ray helmet structure. It expanded steadily and eventually faded into the diffuse corona. A similar effect is observed for the X-ray structures of some active regions detected at the solar limb; they seem to expand more or less continuously, carrying mass and magnetic field outward. This image was taken on 25 January 1992 with the Soft X-ray Telescope (SXT) aboard the Yohkoh satellite of the Japanese Institute of Space and Astronautical Science (ISAS). The SXT was prepared by the Lockheed Palo Alto Research Laboratory (LPARL), the National Astronomical Observatory of Japan, and the University of Tokyo with the support of NASA and ISAS. (Courtesy of LPARL and NASA)

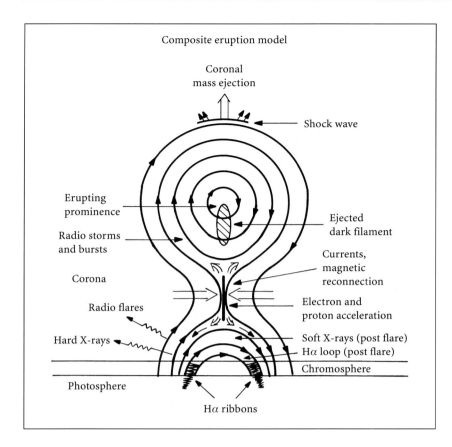

Fig. 7.17. Composite Eruption Model. In this cross-sectional view, a coronal mass ejection, and the erupting prominence that follows it, blast an open pathway into previously closed magnetic fields. Energetic electrons accelerated at the reconnection site give rise to intense, impulsive radio and hard X-ray radiation. When the reconnected magnetism regroups to form closed structures, post-flare loops shine at soft X-ray and hydrogen-alpha wavelengths.

Yet, while we believe magnetism to be the ultimate source of energy involved in both solar flares and coronal mass ejections, no one has ever measured the predicted depletion of magnetic energy that supposedly spawns eruptive outbursts on the Sun. Perhaps the instruments are not sensitive enough, or maybe all the magnetic action occurs in the unseen corona. (Magnetographs only measure the magnetic field in the underlying photosphere.) The available magnetic energy might greatly exceed the amount released during a solar eruption, so little overall change in the magnetism would be observed, but if this is the case why aren't the eruptions more powerful and why don't they occur more frequently?

Whatever the explanation, we have no direct observational evidence that stored magnetic energy powers eruptions on the Sun. Moreover, even if current-carrying magnetism does supply the energy, the exact mechanism of releasing that energy and converting it into heating the gas and the acceleration of particles remains unknown.

It has been supposed that the energy required for eruptions is stored in stressed magnetic structures, and that the magnetic fields rearrange themselves into a simpler configuration after the event. Solar flares do, in fact, occur in regions of strong magnetic shear in the photosphere. However, many sheared regions never erupt, so contorted magnetism seems to be a necessary but not sufficient condition for solar eruption.

Debates continue to rage over exactly what strikes the match that ignites explosions on the Sun. Magnetic fields coiled up in the interior could bob into the corona and interact with preexisting ones, or existing coronal loops may be brought into contact by the shearing and twisting motion of their photospheric footpoints. The eruptions might even be triggered by the disappearance of coronal loops when they cancel out and return back inside the Sun. It could be the emergence, interaction or submergence of coronal loops, but no one knows for sure.

Thus, the Sun's sudden and unexpected outbursts remain as unpredictable as most human passions. They just keep on happening, and even seem to be necessary to purge the Sun of pent-up frustration and to relieve it of twisted, contorted magnetism. As we shall next see, this erratic, unpredictable, impulsive behavior of the Sun is of enormous practical interest to us humans on Earth.

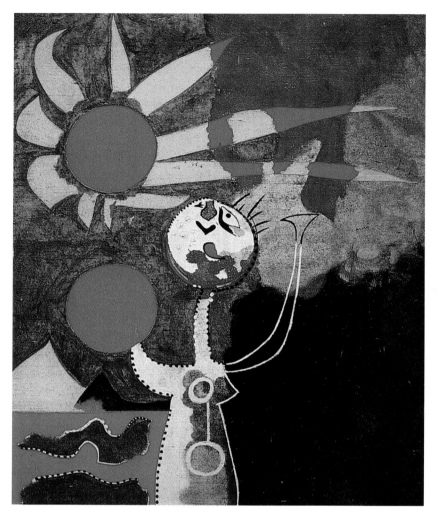

Woman in Front of the Sun. 1938.
In this painting by Joan Miró, a bright Sun sends out rays of hope in the darkness of night. A tall, erect woman is trying to touch the bright red Sun, while holding a red ball or star in another hand. Many of Miró's paintings present terrestrial forms against the powerful red Sun, as well as the stars of the night sky that gave him so much comfort and inspiration. (Private Collection)

CHAPTER 8

Energizing Space

8.1 THE INGREDIENTS OF SPACE

Several decades ago, before the space age, we thought that an imaginary boundary separates the Earth's thin, life-sustaining atmosphere from an empty wasteland called outer space. Back then, we visualized our planet as a solitary sphere travelling in a cold, dark vacuum around the Sun, warmed to life by its radiation. But we now know that the space outside the Earth's atmosphere is not empty! The Earth is immersed in a vast and shifting web of subatomic particles and magnetic fields that come from the Sun.

So, light and heat are not the Sun's only contribution to our environment. The space between the planets, once thought to be an utter void, is swarming with hot, invisible pieces of the Sun. This active, ever-changing maelstrom originates on and flows outward from the Sun. It is an electrified gas or plasma made up of more or less equal numbers of positively charged hydrogen nuclei, or protons, and negatively charged free electrons. We on the surface of the Earth are protected and shielded from these invisible, subatomic particles by our atmosphere and the Earth's own huge magnetic field, and are therefore normally unaware of them.

Thus, one ingredient of space is a plasma. It is a fourth state of matter distinguishable from the more familiar gaseous, liquid, and solid ones. A plasma forms when electrons are stripped from gaseous atoms, leaving an electrically neutral collection of electrons and ions that are free to move about. (Ions are atoms that are missing one or more electrons, so they are positively charged; electrons have a negative charge, and an un-ionized atom is neutral without a charge.) The plasma state of matter can conduct electricity and interact strongly with electric and magnetic fields. The Sun and other stars, and indeed most of the Universe is in the plasma state.

Invisible magnetic fields are another ingredient of space. They emanate from the Earth, as well as the Sun. As early as 1600, William Gilbert, physician to Queen Elizabeth I of England, published a treatise, *De magnete: magnus magnes ipse est globus terrestris*, translated "About magnetism: the terrestrial globe is itself a great magnet". As suggested by Gilbert, the Earth has a huge dipolar magnetic field with two poles, north and south. It's as if there were a bar magnet at the center of the Earth,

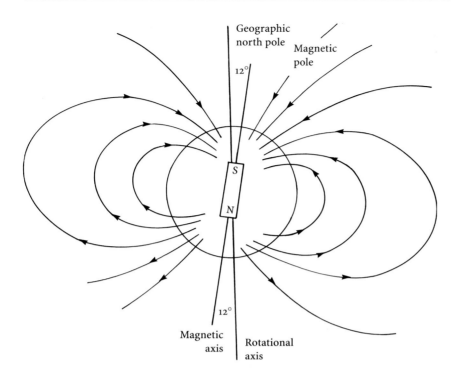

Geographic
north pole
Magnetic
pole
12°
S
N
12°
Magnetic
axis
Rotational
axis

Fig. 8.1. Dipole Field of Earth. The Earth's magnetic field looks like that which would be produced by a bar magnet at the center of the Earth, with the lines of force looping out of the south pole and into the north pole. The magnetic axis is tilted at an angle of 12 degrees with respect to the Earth's rotational axis. This dipolar (two poles) configuration applies near the surface of the Earth, but further out the magnetic field is distorted by the solar wind.

with lines of magnetic force emerging from the south pole, looping out through the space near Earth, and re-entering at the north pole (Fig. 8.1). Since the geographic poles are located near the magnetic ones, compass needles point north-south, within about 12 degrees, which is just what Gilbert intended to explain.

Interplanetary space also contains a small number of very energetic charged particles of both cosmic and solar origin. Streams of energized particles, called cosmic rays, travel between the stars, bringing traces of explosive stellar death (see Focus 8A). Powerful solar eruptions also hurl very energetic charged particles into space.

So, other than the cosmic rays and the byproducts of temporary solar eruptions, the main ingredients of the invisible space between the planets are an electrified plasma and magnetic field. Emanating from the Sun, this continuous energy-laden stream behaves as a wind, a solar wind, that interacts with the Earth's own magnetic environment.

8.2 PROBING THE SOLAR WIND

The entire solar system is bathed in a hot, low-density, high-speed gale that continuously blows from the Sun and carries its own weak magnetic field along with it. This solar wind, which is described in Sects. 6.5, 6.6 and 6.7, is the hot, million-degree outermost layer of the solar atmosphere, called the corona, expanding into cold, vacuous interstellar space. We know the solar wind is always out there, but we cannot see it or feel it.

Only by launching a spacecraft can we directly measure *in situ* the invisible constituents of the solar wind. Spacecraft carry instruments above the absorbing, distorting atmosphere and deflecting magnetic fields of the Earth, studying the charged subatomic particles and electric and magnetic fields where they are found, both within the space near the Earth and further out in the solar wind.

Pouring out of the Sun at about a million tons a second, the solar wind has been diluted to a rarefied plasma by the time it reaches the Earth. Space probes have also shown that the magnetism entrained in the solar wind has been dragged, stretched, and enormously weakened by the time it reaches the Earth's orbit. However, the Sun's wind rushes on with little diminution in speed, for there is almost nothing out there to slow it down.

8.3 EARTH'S MAGNETIC COCOON

Invisible powers collide, sometimes violently, in the vast reach of space between the Earth and the Sun. There the hot, high-speed, magnetized solar wind, which continuously blows from the Sun and rushes toward the Earth, meets the Earth's magnetic field, generated as in a dynamo by electric currents deep within the Earth's core. (This turbulent encounter occurs fairly close to home, usually at a distance of about ten times the Earth's radius.)

FOCUS 8A
Cosmic Rays

Cosmic rays are extraordinarily energetic elementary particles and atomic nuclei that rain down on the Earth from all directions, travelling at nearly the speed of light. They form a third ingredient of space within our solar system in addition to the electrified plasma and magnetic fields that emanate from the Sun. Cosmic rays were discovered in 1912 when Victor Franz Hess took a Geiger counter, used to study energetic particles emitted by radioactive atoms like uranium, on a balloon flight. The measured radioactivity at first decreased with altitude, as would be expected from atmospheric absorption of particles emitted by radioactive rocks, but the Geiger counter's signal unexpectedly increased at even higher distances above the Earth's surface. This meant that the signals were of extraterrestrial origin. By flying his balloon at night and during a solar eclipse, when the high-altitude signals persisted, Hess showed that they could not come from the Sun, but from some other unknown cosmic source.

In 1936, more than two decades after his discovery, Hess received the Nobel Prize in Physics for this remarkable feat. By that time cosmic rays had played an important role in deciphering the inner, invisible subatomic world. The cosmic rays enter the Earth's atmosphere with energies greater than could be produced by atom smashers on Earth. They collide with the atoms in the air, shattering them into subatomic pieces. Investigations of these showering byproducts in the early 1930s led Carl David Anderson to confirm the predicted existence of the positron (antiparticle of the electron), for which he was awarded the Nobel Prize in Physics in the same year as Hess.

The new extraterrestrial signals that Hess discovered were called cosmic rays, distinguishing them from the alpha and beta rays emitted during radioactive decay on the Earth; the alpha and beta rays are now respectively known to be helium nuclei and electrons. The cosmic rays consist mainly of the nuclei of hydrogen atoms, or protons, as well as electrons and the nuclei of heavy atoms increasing in weight from helium (two protons) to uranium (92 protons). They provide rare direct samples of matter from outside the solar system and, although few in number, carry with them the story of the more energetic processes in our Galaxy.

Unlike electromagnetic radiation, the charged cosmic ray particles are deflected and change direction during numerous encounters with the interstellar magnetic field that wends its way between the stars. By the time they reach the Earth, cosmic rays have therefore lost their orientation. So, we cannot look back along their incoming path and tell where cosmic rays originated; the direction of arrival just shows where they last changed course. We can only speculate that cosmic rays were accelerated to their tremendous energies by shocks associated with the explosions of distant dying stars, called supernovae.

Cosmic rays are also deflected by the Sun's magnetism that is carried into interplanetary space by the solar wind. As first shown by Scott Forbush in 1954, the number of cosmic rays arriving at Earth therefore varies with the 11-year cycle of solar activity, but in the opposite direction to that expected if cosmic rays came from the Sun. At times of maximum solar activity, more charged cosmic-ray particles are deflected away from the Earth by the Sun's magnetic field that stretches out into interplanetary space. Less extensive interplanetary magnetism during a minimum in solar activity lowers the barrier to cosmic particles inbound from the depths of space, and allows more of them to arrive at the Earth.

Spacecraft have shown that the terrestrial magnetic field deflects the solar wind away from the Earth and hollows out a cavity in it, forming a protective "cocoon" around the planet. This magnetic cocoon, called the magnetosphere, shields the Earth from the full force of the solar wind, and protects our planet from possibly lethal energetic solar particles. Even though the Earth's surface magnetism is several times weaker than a toy magnet, it is still strong enough to block and divert the flow of the tenuous solar wind, like a rock in a fast flowing stream of water or a building in a strong wind on Earth.

The magnetosphere of the Earth, or of any other planet, is that region surrounding the planet in which its magnetic field has a controlling influence on, or dominates, the motions of energetic charged particles such as electrons, protons, or other ions. The basic idea was put forth about a century ago by Kristian Birkeland, who wrote in 1896 that the

> Earth's magnetism will cause there to be a cavity around the Earth in which the [solar wind] corpuscles are, so to speak, swept away.[32]

Birkeland proposed that electrons from the Sun are directed by the Earth's magnetism to the polar regions where they produce auroras, demonstrating his idea in experiments with electrons and magnetic spheres (see Sect. 8.7). The term magnetosphere was coined by Thomas Gold in 1959, more than half a century later.

Space probes have now encountered every planet except Pluto, showing that six of them are wrapped in their own magnetic sheathing. Mer-

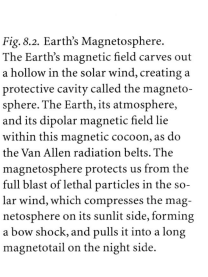

Fig. 8.2. Earth's Magnetosphere. The Earth's magnetic field carves out a hollow in the solar wind, creating a protective cavity called the magnetosphere. The Earth, its atmosphere, and its dipolar magnetic field lie within this magnetic cocoon, as do the Van Allen radiation belts. The magnetosphere protects us from the full blast of lethal particles in the solar wind, which compresses the magnetosphere on its sunlit side, forming a bow shock, and pulls it into a long magnetotail on the night side.

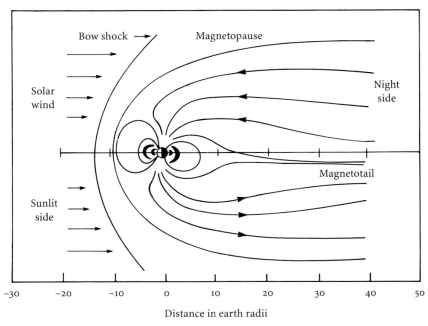

cury, Earth, Jupiter, Saturn, Uranus, and Neptune, have magnetic fields of sufficient strength to deflect the solar wind and form a comet-shaped cavity called a magnetosphere; Venus and Mars have no detectable magnetism.

Near the Earth, the magnetic field retains its dipolar configuration, bulging outward into space near the equator and converging inward at its two poles, like the pattern of iron filings scattered near an ordinary bar magnet (also see Fig. 8.1). Terrestrial magnetism weakens as it extends to greater distances, eventually becoming distorted by the Sun's wind. Far from the Earth, the term sphere in magnetosphere loses its strict geometrical meaning, and instead implies a more general sphere of influence. Out there, the solar wind takes over and continuously molds the Earth's magnetosphere into a changing, asymmetric shape (Figs. 8.2 and 8.7). If it could be seen, the outer boundary of the magnetosphere would have an appearance not unlike that of a comet.

Rarefied as it is, the solar wind still possesses the power to bend and move things in its path. It compresses and flattens the outer boundary of the magnetosphere into a blunt-nosed shape on the dayside facing the Sun; on the nightside of the Earth away from the Sun, the terrestrial magnetic field is stretched out and turned inward upon itself by the relentless solar wind, forming an invisible magnetotail that always points downwind like a weather vane (Fig. 8.7). The magnetotail consists of two lobes of oppositely-directed magnetic field lines derived from the two hemispheres of the terrestrial magnetic field. Magnetic field lines emerging from the southern hemisphere are directed away from the Sun, those in the northern lobe are directed sunward.

A shock wave forms in front of the Earth's magnetosphere on the sunward dayside; it is called a bow shock because it is shaped like the waves that pile up ahead of the bow of a moving ship (Fig. 8.3). At the bow shock the solar wind abruptly decelerates to subsonic speed and heats up, like the wheels of a car slamming on its brakes or an ocean wave crashing into foam at the shore. Most of the solar wind is then deflected around the Earth, but some of its particles are reflected back from the bow shock into the onrushing solar wind, like the eddies around a rock in a river.

The magnetosphere is constantly being buffeted by the variable solar wind. Monitors in space show that the solar gale is never steady. Like winds on Earth, it is punctuated by gusts and has its own tempestuous weather.

When the solar wind pressure is high, the bow shock moves inward, and when the pressure drops, the Earth's magnetic domain expands; it's like squeezing a rubber ball and letting it go. The entire magnetosphere compresses and expands, changing size constantly as the solar wind varies in density and speed. These variations are frequently caused by violent eruptions on the Sun; they occur more often during the maximum in the 11-year cycle of solar activity.

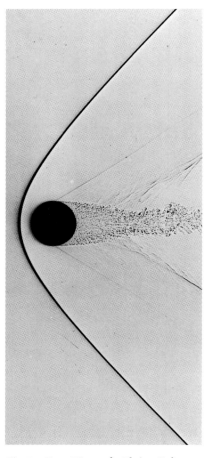

Fig. 8.3. Bow Wave of a Flying Sphere. A small sphere, in free flight through the air, forms a bow wave that is similar to the bow shock formed by the interaction of the solar wind with the Earth's magnetic field (see Fig. 8.2). The flying sphere also has a turbulent wake that may resemble the Earth's magnetotail. (Courtesy of Alexander C. Charters, Marine Science Institute, University of California, Santa Barbara. Made at U. S. Army Ballistics Research Laboratory, Aberdeen Proving Ground, Maryland)

Fig. 8.4. Invisible Worlds. The Sun is enveloped by an invisible, low-density electrified gas, called the solar wind, that continuously expands out past the planets to the very edge of the solar system. It creates the heliosphere – the sphere of influence of the Sun (from *helios* the Greek word for the Sun). The magnetic field of the Earth hollows out a protective cavity in the solar wind, called the magnetosphere, that is buffeted by the solar wind on the sunlit side and drawn out by the solar wind into the shape of a teardrop on the night side. The invisible magnetosphere is that region surrounding the Earth where its magnetic field controls the motions of charged particles. (Courtesy of NASA)

8.4 PENETRATING EARTH'S MAGNETIC DEFENSE

Particles in the solar wind transport only one ten-billionth the energy of that carried by sunlight, and Earth is protected from the full blast of the dilute, varying solar wind by the terrestrial magnetosphere. The Earth's magnetic shield is so perfect that only 0.1 percent of the mass of the solar wind that hits it manages to penetrate inside. Yet, even that small fraction of the wind particles has a profound influence on the Earth's environment in space. They create an invisible world of energetic particles and electric currents that encircle the Earth in space (Figs. 8.4 and 8.5).

Solar-wind particles being carried past the Earth penetrate its magnetic defense, primarily along solar magnetic field lines that connect to the Earth's magnetic field. The merging is most effective if the two fields point in opposite directions when they meet (Fig. 8.6). This magnetic coupling occurs on the dayside facing the Sun and is dragged downstream all along the length of the Earth's magnetotail. This immense magnetic tail forms the bulk of the magnetosphere, and therefore provides the main location for breaching the Earth's magnetic defense.

Fig. 8.5. Sun-Earth Relations. Two invisible powers collide in the vast reaches of space between the Earth and the Sun, where the hot, expanding corona, or solar wind, meets the Earth's magnetic field that deflects the electrified solar wind, shielding us from possibly lethal solar particles. Energetic solar activity, such as solar flares or coronal mass ejections, produces gusts in the solar wind that buffet and distort the terrestrial magnetosphere, enabling energetic charged particles to penetrate the Earth's magnetic defense. The Sun thereby feeds an unseen world of high-speed particles and magnetic fields that encircle the Earth in space, including the donut-shaped radiation belts, while also lighting up the polar auroras and endangering astronauts and satellites. (Courtesy of NASA)

1 Hydrogen nuclei
2 Solar core
3 Helium
4 Sunspots
5 Sun
6 Magnetic loops
7 Solar flare
8 Solar wind
9 Energetic charged particles
10 Corona
11 Coronal mass ejection
12 Terrestrial magnetosphere
13 Auroral zone
14 Radiation belts
15 Magnetotail

The passing solar wind is slowed down by the connected fields and decelerates in the vicinity of the tail. Energy is thereby extracted from the nearby solar wind and drives a large-scale circulation, or convection, of charged particles within the magnetosphere (Fig. 8.7). While creating and sustaining the magnetotail, the solar wind brings the oppositely-directed tail lobes into close contact, where they can merge together and send material back toward the Earth. On average, the merging within the tail balances the dayside merging to complete the circulation pattern.

Fig. 8.6. Coupling between Sun and Earth. Most of the solar wind flows around the Earth, but some of the wind's energetic particles penetrate the Earth's magnetic defense. (Courtesy of NASA)

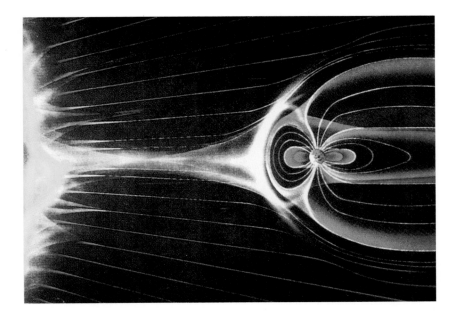

Magnetic connection within the tail does not proceed continuously, however, but rather in a series of disturbances or spurts, called substorms, that release pent-up tension and return the magnetosphere to a more stable, low-energy state. These substorms last about one hour and occur typically a few times per day.

The tension arises and substorms begin when the dayside merging rate increases owing to the magnetic field in the solar wind turning in the opposite direction to the terrestrial magnetism. The volume of field

Fig. 8.7. Plasma Circulation. Open arrows indicate the circulating flow of plasma in this cross-section of the Earth's magnetosphere. Plasma in the solar wind is deflected at the bow shock (*left*), flows along the magnetopause into the magnetic tail (*right*), and is then injected back toward the Earth and Sun within the plasma sheet (*center*). This plasma circulation pattern is driven by magnetic tension created when the magnetic field in the solar wind merges with that of the Earth. This mechanism may provide 80 to 90 percent of the coupling between the solar wind and the Earth's magnetosphere, but solar-wind plasma may also enter into other weak points in the Earth's magnetism such as the turbulent bow shock and the polar cusps. (Courtesy of Tom Hill, Rice University)

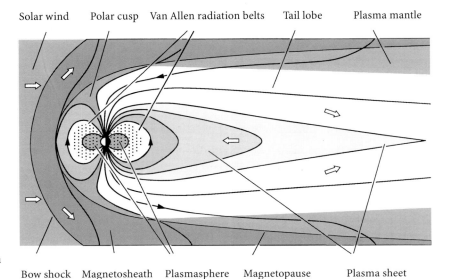

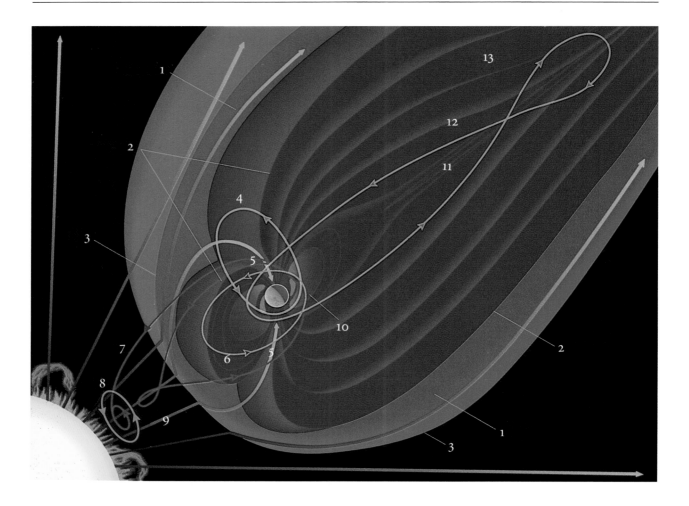

lines in the tail lobes then increases until a substorm occurs. This explosively releases magnetic energy stored in the tail, much like the process
that powers some solar flares (see Sect. 7.4). The magnetotail then snaps
like a rubber band that has been stretched too far. The snap catapults
part of the tail downstream into space, creating a gust-like eddy in the
solar wind.

The other part of the tail, propelled by energy released in the magnetic merging, rebounds back toward the Earth. Electrons and ions hur-
tle along magnetic conduits that are connected to the Earth, linking the
solar wind to both equatorial storage regions and down into the polar
caps. The electrons that are guided into the polar regions augment and
intensify the northern and southern lights, or auroras (see Sect. 8.7). The
tail continues to reform into another, potentially explosive, unstable configuration to await the next substorm. They continue until the magnetic
field in the solar wind changes direction and the dayside merging is
closed down or decreases.

◁ *Fig. 8.8.* Inside the Magnetosphere.
A bow shock upstream marks the impact of the solar wind on the Earth's
magnetic shield; this boundary moves
in and out on the sunward side depending on the wind's strength. The
magnetosheath is a turbulent layer
immediately behind the shock front
in which electrons and protons reach
energies several thousand times as
great as in the incoming solar wind.
The magnetopause is the outer boundary of the magnetosphere. On the
sunward side, the magnetic field
opens up into two funnel-shaped
polar cusps; solar wind particles can
travel through the cusps to the Earth's
upper atmosphere. Energetic wind
particles can enter the magnetosphere at other weak points, including
the bow shock and the magnetotail.
The Van Allen belts are donut-shaped
regions encircling the Earth, populated by energetic electrons and protons.
The orbits of spacecraft designed to
investigate crucial areas of the magnetosphere and its interaction with
the solar wind are also shown – also
see Fig. 8.17. (Courtesy of NASA)

1   Magnetosheath region
2   Magnetopause
3   Bow shock
4   Polar Spacecraft
5   Cusp
6   Cluster Spacecraft
7   Wind Spacecraft
8   SOHO Spacecraft
9   Solar wind
10 Radiation belts
11 Neutral sheet
12 Geotail Spacecraft
13 Magnetotail region

The inside of the magnetosphere is therefore a highly dynamic,
interactive and complex environment that depends on the ever-changing
Sun (Fig. 8.8). Magnetic fields that are conducive to dayside merging can
be brought by coronal mass ejections, which occur at times of increased
solar activity during the maximum of the 11-year sunspot cycle. These
transient eruptions on the Sun thus set into motion the substorm cycle of
magnetic storage and release.

8.5 STORING INVISIBLE PARTICLES WITHIN EARTH'S MAGNETOSPHERE

Charged particles flowing from the Sun can enter the Earth's magnetic
domain and become trapped within it. The particles that supply the
aurora are stored on the stretched dipolar field lines, earthward of the tail
magnetic connection site. This storage region is called the plasma sheet
(see Fig. 8.7). It acts as a holding tank of electrons and ions, suddenly releasing them when stimulated by the ever-changing Sun. Nearer the
Earth particles are stored in the radiation belts, regions of unexpectedly
high flux of high-energy electrons and protons that girdle the Earth far
above the atmosphere in the equatorial regions.

Inspired by Birkeland's experiments with electrons and a dipolar
magnetic sphere (also see Sect. 8.7), Carl Störmer, a young theoretical
physicist in Oslo, studied mathematically the motion of charged particles in the magnetic field of a dipole. Using tedious numerical calculations before the computer age, he showed in 1907 that electrically
charged particles from the Sun would be captured by Earth's dipolar
magnetism and forced into magnetic reservoirs near the planet, spiraling around the magnetic field lines and bouncing back and forth
between the Earth's magnetic poles for long periods (Fig. 8.9). The spirals described by the trapped particles tighten up in the stronger magnetic field close to the Earth, until they are deflected at a mirror point
near one of the poles. Still, most scientists were completely surprised by
the unexpected, space-age discovery of toroidal or donut-shaped belts of
electrons and protons trapped within the terrestrial magnetosphere.

The first American satellite Explorer 1, launched in January 1958 in
response to the Soviet Union's successful Sputnik satellites in October
and November 1957, had a Geiger counter aboard to survey the intensity
of cosmic rays. The instrument recorded the expected cosmic rays near
the Earth, but unexpectedly detected none at higher altitudes. The effect
was confirmed two months later by Explorer 3. (Explorer 2 went into the
ocean.) It turned out that the satellites had entered a dense region of energetic particles that saturated the Geiger tube, causing the counter to
read zero. It actually showed that space is filled with "radiation" of intensities a thousand times greater than expected, suggesting that energetic,

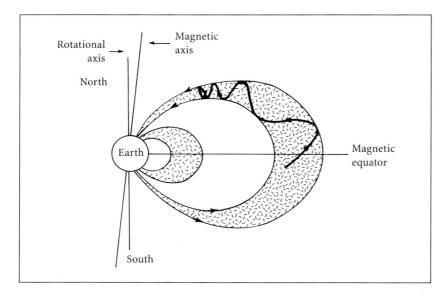

Fig. 8.9. Van Allen Belts. The first scientific spacecraft, Explorer 1 launched in 1958, and Pioneer III launched in the early 1960s, revealed two distinct bands of high-energy electrons and ions girdling the Earth above the equator. These regions are now called the inner and outer Van Allen radiation belts, named after Dr. James Van Allen who first observed them. Here we show the path of a charged particle in one of the belts; it is a trapped prisoner of the Earth's magnetic field. As the energetic particle approaches the polar regions, the increasing magnetic intensity turns it around; the particle then spirals in just a few seconds to a similar turning point at the opposite pole. The energetic particles trapped in the Van Allen belts can damage the microcircuits, solar arrays and other materials of spacecraft that pass through them.

charged particles encircle the Earth within donut-shaped regions near the magnetic equator.

Instruments on board Pioneer 3, an unmanned American spacecraft launched to the Moon in late 1958, discovered a larger, more remote donut-shaped reservoir of energetic, charged particles. These regions are sometimes called the inner and outer Van Allen radiation belts, named after James A. Van Allen, the scientist who first observed them; they are always dubbed radiation belts since the charged particles that they contain were known as corpuscular radiation at the time of their discovery.

The inner radiation belt consists mostly of highly energetic protons (10 to 700 MeV, but those in the higher energy range are rare) and is located in the Earth's magnetic equatorial region at an average distance of 1.5 terrestrial radius above the equator; the outer radiation belt contains mainly high-speed trapped electrons (a few MeV) from the Sun, girdling the Earth around the equator at about 3.0 terrestrial radii above the Earth's surface (Fig. 8.9). Both of these belts are in turn enclosed within the magnetosphere – the cavity that the Earth's magnetic field hollows out in the solar wind.

Van Allen and his colleagues at the University of Iowa were fully aware of Störmer's prior work, reporting in 1959 that:

> The existence of a high intensity of corpuscular radiation in the vicinity of the Earth was discovered by apparatus carried by satellite 1958 alpha [Explorer 1] ..... It was proposed in our May 1, 1958 report that the radiation was corpuscular in nature, was presumably trapped in Störmer-Treiman lunes about the Earth,

and was likely intimately related to that responsible for aurorae. On the basis of these tentative beliefs it was thought likely that the observed trapped radiation had originally come from the Sun in the form of ionized gas.[33]

Most of the charged particles found in the radiation belts do, in fact, originate in the solar wind. They enter the magnetosphere and circulate within it (see Fig. 8.7), but the circulation rate changes in response to changes in the coupling between the solar-wind magnetic field and the terrestrial one. Strong circulation brings the particles close to Earth, and when the circulation weakens they get left behind and become trapped in the radiation belts. Other, low-energy particles come from the Earth's upper atmosphere, or ionosphere, beneath the radiation belts.

Satellite-borne instruments have recently detected a third radiation belt that contains high-energy nuclei, called anomalous cosmic rays, that originate outside the solar system (Fig. 8.10). Unlike the other two radiation belts, which are mainly fed by electrons and protons from the Sun, the new one contains interstellar material that was once hurled into space by dying stars or during the explosive origin of the Universe. It includes ions of nitrogen, oxygen and neon, that could only have been made inside stars other than the Sun, as well as ions of helium that may have originated in the big-bang explosion that gave rise to the expanding Universe. (The anomalous cosmic rays have lower velocities and energies than other cosmic rays, and consist of atoms that have lost only one or two of their outer electrons, rather than most of them, presumably resulting from a relatively nearby, less-violent process than that responsible for most cosmic rays that travel at nearly the speed of light.)

Fig. 8.10. New Radiation Belt. A torus of high-energy particles creates a third radiation belt (*yellow*) alongside the familiar Van Allen belts. The outer Van Allen belt (*purple*) consists mainly of electrons from the Sun; the inner one (*blue*) contains energetic solar protons. The newly discovered belt contains heavy nuclear ions that originated outside our solar system and once drifted between the stars. The cosmic ions, as well as the solar electrons and protons, are trapped by the Earth's magnetic field and bounce back and forth between its magnetic poles. (Ian Worpole, © 1993 Discover Magazine)

1 Outer Van Allen belt
2 Inner Van Allen belt
3 Magnetic field lines
4 New belt

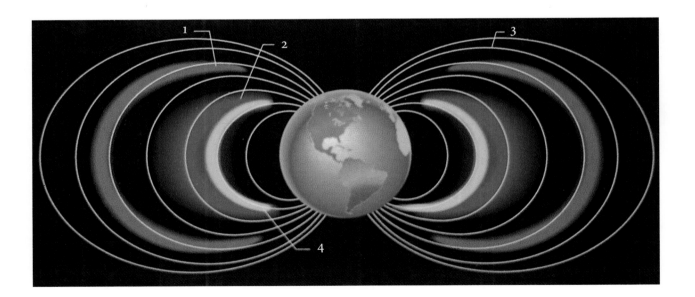

The trapped ions are up to a thousand times more abundant than other cosmic rays (see Focus 8A), and are hence of peculiar, anomalous composition. They share space with the inner solar-proton radiation belt. The anomalous cosmic ray belt was described in detail in 1993 using data from NASA's Solar, Anomalous, and Magnetospheric Particle Explorer (SAMPEX), confirming less accurate results obtained in 1991 by scientists using the Russian COSMOS satellites.

The anomalous cosmic rays trapped in the new radiation belt probably increase in number as the Sun's activity drops during its 11-year cycle, when fewer of them are deflected by magnetic fields extending from the Sun into interplanetary space. Investigations by Pioneer 10 and 11, as well as Voyager 1 and 2, indicated that the intensity of anomalous cosmic rays increased when these spacecraft moved away from the Sun past the distant giant planets and toward interstellar space, suggesting an origin outside the solar system.

It is thought that these ions have entered the new radiation belt after a complex voyage, arising from electrically neutral atoms that originate in the thin gas between the stars. When interstellar atoms enter the solar system, they can become electrically charged by losing one or more electrons (partially ionized) as the result of solar ultraviolet radiation or collisions with solar-wind particles. These ions are then swept to the edge of the solar system by the solar wind, and they are accelerated to cosmic ray energy levels when they encounter the shock wave out there (also see Sect. 6.7). Some of the energized ions then make their way back to the Earth as anomalous cosmic rays, where they are trapped by the dipolar, terrestrial magnetism within the newly-discovered belt. They therefore provide a direct sample of the interstellar medium around our solar system.

8.6 ACTIVE EXPERIMENTS IN SPACE

One difficulty with studying outer space is that passive observations are usually employed, either by *in situ* sampling with spacecraft or by remote sensing from the ground. Space is normally not subject to the sort of hands-on experiments carried out in a terrestrial laboratory. This difficulty has been overcome by actively stimulating Earth's environment in space and measuring its response.

Scientists use spacecraft to alter the space near Earth under controlled conditions. An example in the 1990s is a joint NASA and Department of Defense venture dubbed CRRES for the Combined Release (CR) and Radiation Effects (RE) Satellite (S) Program. The CRRES program comprises a high-altitude, orbiting satellite and low-altitude, non-orbiting sounding rockets that release vaporized chemicals and other substances in space. The released material includes a group of elements

Fig. 8.11. Chemicals Light Up Space. A yellow-green ball of barium atoms shines by reflected sunlight, becoming bigger and brighter than a full Moon. The neutral, or un-ionized, barium atoms are not affected by magnetic fields, and expand into a sphere that moves with the same speed as the NASA Combined Release and Radiation Effects Satellite (CRRES) that released them. The electrons of other barium atoms have been knocked off by ultraviolet sunlight, thereby creating ions with a positive charge. Once the barium becomes ionized, it also emits light but at a different wavelength with a blue color. The barium ions line up along the magnetic field lines, creating a long blue wake and illuminating the otherwise-invisible magnetic structure. A similar thing happens to comets that are enveloped by a spherical halo of neutral hydrogen atoms, and contain long ionized tails that can be snapped off by the interplanetary magnetic field. (Courtesy of Morris Pongratz, Los Alamos National Laboratory, who took this picture from a site in St. Croix using a standard 35 mm camera)

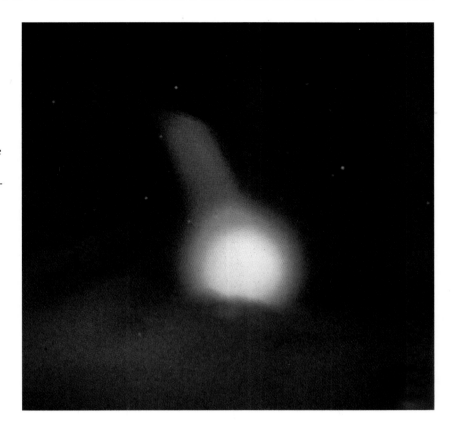

known as alkaline-earth metals (familiar elements such as sodium, barium, calcium and lithium).

These chemical elements have two useful properties when they are released in space in vapor form. The first is that in the presence of ultraviolet sunlight they turn from neutral atoms to positively charged ions, but at varying speeds for different elements. The second is that they also glow. They absorb the Sun's light and re-emit it at specific colors that depend on the element, forming brilliant colored clouds visible from much of the Earth's surface.

The chemical vapors overcome a major obstacle in studying space by lighting it up and making it visible (Fig. 8.11). Sunlight passes right through the transparent space near Earth with no noticeable effect, so outer space is usually invisible.

The glowing charged material can be used to trace out otherwise-invisible magnetism in space. The magnetic fields present a barrier to charged particles, and provide a conduit that guides their motion. The colored ions therefore streak out along the magnetic field and outline its shape. In contrast, neutral, or un-ionized, atoms, reflect sunlight, and create a colored, spherical cloud that moves through space with the same speed as the satellite that released it.

The injection of these substances into space does not pose a threat even though some of the materials such as sodium, barium and lithium are toxic or dangerous when in contact with humans. They are released well above the life-supporting atmosphere and the amounts are so small that any chemicals which drift down would be dispersed over an enormous area in concentrations that are much less than those resulting from human activities such as burning coal (which releases barium into the atmosphere).

In another example of actively perturbing space, astronauts have fired beams of high-speed electrons into the atmosphere, like the light sabers activated by imaginary characters on Star Wars. The high-speed electrons make the air glow in green and red colors. A similar process creates the auroras, or the northern and southern lights, with the Sun providing the electrons in a complicated path that includes acceleration in the magnetosphere (see Sect. 8.7).

Nuclear weapons were once used to disturb space by means of high-altitude detonations. Scientists suggested exploding a bomb within the inner radiation belt to study the auroral effects of the newly injected particles, and the military correctly supposed that the injected material would destroy orbiting satellites and produce a prolonged radio communications blackout. Between 1958 and 1962 atomic bombs were exploded in the atmosphere by the United States military, producing brilliant artificial auroras, but spreading so much damaging radioactive fallout throughout the world that atmospheric nuclear tests have now been banned by international agreement.

In retrospect, the entire nuclear test episode is hard to believe. Yet it is perhaps not so amazing as the suggestion by the chemist Harold Clayton Urey to send an atomic bomb to the Moon and blow it up, with the hope that parts of the Moon would fall back to Earth where they could be recovered. In this Nobel-prize winning scientist's own incredible words, written in 1956:

> One peaceful use of the military employment of atomic energy and intercontinental ballistics would be the delivery of an atomic bomb to the Moon's surface with the expectation that some material would leave its surface and arrive later, intact and unheated, at the Earth's surface.[34]

Peaceful uses of atomic bombs indeed. Fortunately, the United States subsequently decided to send men to the Moon instead, safely returning lunar samples to find out what the Moon is made of without further disturbing the Universe.

Still, even today we continue to perform experiments in space that have rather startling effects. When ordinary, everyday water is dumped into the upper atmosphere, it neutralizes the ions that are there, literally creating a hole in the sky. Scientists have also burned holes in this part of our atmosphere by pumping a lot of electromagnetic energy into it at just the right frequency, thereby reinforcing the natural quivering of the ionized gas until the vibrations go out of control.

Thus, material injections, beams of accelerated particles, bombs, and electromagnetic waves have all been used to actively probe the regions that envelop the Earth. Instruments measure how the disturbance propagates, grows and decays, tracing out the ordinarily unseen properties of space. The artificially-introduced charged particles follow the same magnetic pathways as the electrons that produce the auroras, the only natural, visible manifestation of the interaction of the Sun with the Earth, to which we now turn.

Fig. 8.12. Aurora Borealis. Spectacular curtains of red and green light are found in these photographs of the fluorescent Northern Lights, or Aurora Borealis, taken by Forrest Baldwin in Alaska. (Courtesy of Kathi and Forrest Baldwin, Palmer, Alaska)

8.7 NORTHERN AND SOUTHERN LIGHTS

The northern and southern lights are one of the most magnificent and earliest-known manifestations of the myriad links between the Sun and the Earth. They illuminate the Arctic and Antarctic skies, where curtains of multi-colored light dance and shimmer across the night sky far above the highest clouds (Fig. 8.12). Perhaps because they

imitate the rising morning Sun, they are often called the *aurora* after the Roman goddess of the rosy-fingered dawn; this designation has been traced back to the time of Galileo Galilei (1564 to 1642). The auroras seen near the north and south poles have been given the Latin names aurora borealis, for northern lights, and aurora australis, for southern lights.

Auroral activity is not rare; it is almost always present! The aurora borealis is regularly seen by residents in far northern locations, every clear and dark winter night. (During the summer months of midnight Sun there is little or no darkness, so auroras, though present in the sky, are difficult to see.) Even today, the winter aurora brings solace to circumpolar inhabitants, reminding them of the eventual return of life-sustaining sunlight. However, most people never see the awesome lights, for auroras are normally confined to high latitudes in the north or south polar regions with relatively few inhabitants. In addition, only an intense aurora can be noticed against bright city lights.

The aurora borealis is spectacular and easily seen with the unaided eye, either from remote Arctic locations or during infrequent occasions from more temperate locales. It has therefore been documented for many centuries.

Some of the earliest written accounts of the northern lights are found in Mediterranean countries where schools and libraries flourished long ago. Greek records of the aurora date back to Plutarch, in 467 B.C. Aristotle described an aurora, probably the one occurring in 349 B.C., as being blood-red in color and with chasm-like or trench-like shapes. However, spectacular auroras rarely extend as far south as Greece, perhaps every 50 to 100 years.

Since civilization began, the multi-colored lights have always been most frequently observed at far northern latitudes, so some of the most vivid and comprehensive accounts come from the Scandinavian countries. The oldest written records by Norwegians concerning the northern lights go back to the Viking period (500 to 1300 A.D.). In one Norse chronicle, called *Kongespeilet* or *King's Mirror* and written about 1250 A.D., it is described by:

> These northern lights have this peculiar nature, that the darker the night is, the brighter they seem, and they always appear at night but never by day, most frequently in the densest darkness and rarely by moonlight. In appearance they resemble a vast flame of fire viewed from a great distance. It also looks as if sharp points were shot from this flame up into the sky, they are of uneven height and in constant motion, now one, now another darting highest; and the light appears to blaze like a living flame .... It seems to me not unlikely that the frost and the glaciers have become so powerful there that they are able to radiate forth these flames.[35]

Fig. 8.13. Aurora Australis. The shimmering, colorful aurora is fascinating to watch from below, but the topside view is even better. In this photograph of the Southern Lights, or Aurora Australis, auroral curtains trace terrestrial magnetic field lines which extend upward above the Antarctic. It was taken from the flight deck of the Space Shuttle Discovery during a military mission dedicated to the Department of Defense, in order to study how the aurora might confuse early warning satellites looking for rocket plumes from intercontinental ballistic missiles. Energetic electrons cascading down the magnetic field lines excite oxygen atoms in the atmosphere, causing them to fluoresce and glow with a green color; this airglow occurs at an altitude of 80 to 120 kilometers above the Earth's surface. Less energetic electrons do not penetrate as deeply as the more energetic ones, and cause the oxygen to glow red at altitudes of about 250 kilometers. (Courtesy of NASA)

In his book *The Fram Expedition*, published in 1897, the Norwegian explorer Fridtjof Nansen provided this account, written while he was trapped through the long Arctic winter in the frozen pack ice:

> The glowing fire-masses had divided into glistening, many coloured bands, which were writhing and twisting across the sky both in the south and north. The rays sparkled with the purest, most crystalline rainbow colours, chiefly violet-red or carmine and the clearest green. Most frequently the rays of the arch were red at the ends, and changed higher up into sparkling green .... It was an endless phantasmagoria of sparkling colour, surpassing anything that one can dream. Sometimes the spectacle reached such a climax that one's breath was taken away; one felt that now something extraordinary must happen – at the very least the sky must fall.[36]

Nansen, who won the Nobel Peace Prize in 1922, wrote a few more sentences and could not continue; being thinly dressed and without gloves, he had no feeling left in body or limbs.

Auroras occur with the same frequency and simultaneously in both the southern and northern polar regions of the Earth. Indeed, the two auroras are almost mirror images of each other. Yet, the first recorded sighting of the southern lights has been attributed to Captain Cook on his 1770 voyage of the *Endeavor*. The aurora australis have never achieved a renown comparable to the northern lights, probably because the southern ones are not usually located over inhabited land and are instead seen from oceans that are infrequently travelled.

Nowadays we can use spacecraft to view both the northern and southern lights from above (Figs. 8.13 and 8.14). The Space Shuttle has even flown right through the northern lights. (Visual auroras normally occur between 100 and 250 kilometers above the ground and can extend to about 400 kilometers.) While inside the display, astronauts could close their eyes and see flashes of light caused by the charged auroral particles speeding through their eyeballs. And the view from space is just as magnificent as that from the ground. As the Czechoslovakian astronaut, Vladimir Remek, expressed it:

> Suddenly, before my eyes, something magical occurred. A greenish radiance poured from the Earth directly up to the [space] station, a radiance resembling gigantic phosphorescent organ pipes, whose ends were glowing crimson, and overlapped by waves of swirling green mist.[37]

The play of northern lights in the sky has been described in the folklore of Arctic cultures, where they are often interpreted as the spirits of the dead either fighting or playing in the air. The Vikings thought they represented an eternal battle between the spirits of fallen warriors. Eskimos have described the flickering lights as a dance of the dead, amusing themselves in the absence of light from the Sun, or as signals from the deceased trying to contact their living relatives. An Eskimo word for aurora, *aksarnirq*, translates into *ball player*. For Alaskan Eskimos the spirits are playing ball with the heads of children who dared venture outside during the northern lights.

Fig. 8.14. Aurora from the Shuttle. The aurora appears when solar storms release highly-charged particles that collide with the Earth's upper atmosphere. The ensuing multi-colored walls of shimmering light can best be seen at high latitudes in the Northern or Southern hemispheres. The eerie, beautiful glow of auroras can also be detected from space, as shown in this image of the Aurora Australis or Southern Lights taken from the Space Shuttle Discovery. The colored emission of atomic oxygen extends upward to 200–300 kilometers above the Earth's surface. (Courtesy of NASA)

The flickering, colored lights have also inspired many poets, such as Browning's "belch of fire" as from a "dragon's nostril", Coleridge's "streamy banners of the North" and "a hundred fire-flags sheen", Sir Walter Scott's "spirits riding the northern light", and Wallace Stevens' "polar green color of ice and fire and solitude".

In Norway or Sweden it was a common belief that the northern lights were reflections from silvery shoals of herring who would throw up a flash of light against the clouds when swimming close to the water's surface. According to *The King's Mirror*, the Arctic snow and ice absorb large amounts of light from the long summer midnight Sun and re-radiate or reflect it, like a mirror, as the northern lights in winter time. Almost 400 years later, the French philosopher René Descartes attributed them to sunlight scattered from ice particles found high in the atmosphere at cold northern locations. So, the idea that auroras are reflections from Arctic ice fields or from the cold icy air persisted for centuries, but it was eventually shown to be wrong.

Since auroras become more frequent as one travels north from tropical latitudes, it was thought that they would occur most often at the highest polar latitudes. Early Arctic explorers were therefore surprised to find that their frequency of occurrence did not increase all the way to the poles. In 1860 Elias Loomis, an American professor of natural philosophy, mapped out their geographic distribution, showing that the northern lights form a luminous ring encircling the North Pole. The Swiss engineer and physicist, Herman Fritz, extended Loomis' work, publishing a similar conclusion in 1881 in his then-well-known book *Das Polarlicht*. Loomis and Fritz showed that the intensity and frequency of auroras were greatest in an oval-shaped auroral zone which is about 500 kilometers broad and about 2000 kilometers from the pole. Auroras can occur every night of the year within this zone.

About a century later, the auroral distribution was mapped out in greater detail using all-sky cameras (during the International Geophysical Year in 1957–58). An analysis of hundreds of thousands of photographs, each portraying the sky from horizon to horizon, showed that the auroral zone is an oval-shaped band that is centered on the Earth's magnetic pole.

Today spacecraft look down on the auroral oval from high above the north polar region, showing the northern lights in their entirety. An example is shown in Fig. 8.15, where the glowing oval is projected against a map of the continents. The luminous auroral oval is constantly in motion, expanding toward the equator or contracting toward the pole, and constantly changing in brightness. Such ever-changing auroral ovals are created simultaneously in both hemispheres and can be viewed at the same time from the Moon.

Because the radius of both the oval and the Earth are much larger than the height of the aurora, an observer on the ground sees only a

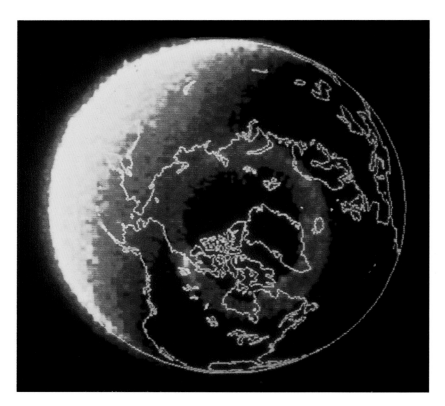

Fig. 8.15. The Auroral Oval. The huge luminous ring of this auroral oval is centered on the north, geomagnetic pole. The oval, some 4500 kilometers across, is due to electromagnetic interactions at the boundary of the Earth's magnetosphere and the solar wind. The bright crescent at the upper left is the illuminated daylight side of the Earth. This image was made by a University of Iowa team using data taken by the satellite Dynamics Explorer 1 at an altitude of 20 000 kilometers. (Courtesy of Louis A. Frank)

small piece of the auroral oval, which can resemble a shimmering luminous curtain hanging down from the cold Arctic sky.

Loomis and Fritz established a general correlation between the occurrence of sunspots and northern lights. When the number of sunspots is large, bright auroras occur more frequently, and when there are few spots on the Sun the intense auroras occur less often. So, the frequency of occurrence of the bright auroras tends to follow the 11-year sunspot cycle of solar activity, suggesting that the Sun somehow controls the brilliance of the northern lights.

The Sun's influence was more fully explained in 1896 when the Norwegian physicist Kristian Birkeland showed that electrons from the Sun could be directed and guided along the Earth's magnetic field lines to the polar regions. As the solar electrons cascade down the magnetic poles into the atmosphere, they are slowed down by collisions with the increasingly dense air, exciting the gaseous atoms and causing them to glow like a cosmic neon sign.

Birkeland demonstrated his theory by sending electrons toward a magnetized sphere, called a terella, using phosphorescent paint to show where electrons struck it. (An electromagnet was placed in the sphere, creating a dipolar magnetic field, and the entire apparatus was placed in a low-density vacuum that represented outer space.) The resulting light indicated that the electrons are curved down toward and around the magnetic poles, and the glowing shapes reproduced many of the observed features of the auroras.

Particle detectors on board rockets launched into auroras in the early 1960s showed that Birkeland was right (at least in part)! The aurora is principally excited by energetic electrons bombarding the upper atmosphere with energies of about 6 keV, or 6000 electron volts, and speeds of about 50 000 kilometers per second. Currents as great as a million amperes can be produced along the auroral oval, and the electric power generated during the discharge is truly awesome – about ten times the annual consumption of electricity in the United States. (Birkeland also proposed the existence of such auroral currents, but this idea was then largely ignored.)

When the electrons slam into the upper atmosphere, they collide with the oxygen and nitrogen atoms there and excite them to energy states unattainable in the denser air below. (The most abundant constituents of our atmosphere are oxygen and nitrogen, respectively comprising 21 and 78 percent, and their energized auroral transitions are "forbidden" at high densities.) The pumped-up atoms quickly give up the energy they acquired from the electrons, emitting a burst of color in a process called fluorescence.

Excited oxygen atoms radiate both green (5557 Angstroms) and red (6300 and 6364 Angstroms) light, but because very energetic electrons are required to make oxygen glow red, the most common auroral color is green. Each color also has a specific altitude range; the green oxygen emission appears at about 110 kilometers and the red oxygen light at 200 to 400 kilometers. The bottom edge of the most brilliant green curtains are sometimes fringed with the pink glow of neutral (un-ionized) molecular nitrogen, and rare blue or violet colors are emitted by ionized nitrogen molecules.

So, the multi-colored auroral light show is caused by energetic electrons colliding with oxygen and nitrogen atoms in the air, but where do the electrons come from and how are they energized? Since the most intense auroras occur at times of maximum solar activity, it was once thought that the auroral electrons are energized during eruptions on the Sun and hurled directly down into the upper atmosphere through the narrow, funnel-shaped polar cusps. In another popular theory, solar wind electrons were supposed to be held within the Van Allen radiation belts before being squirted out into the auroral zones, like a squeezed tube of toothpaste, as the result of excessive solar activity. However, solar particles coming in from the polar route apparently do not have enough energy to make all of the auroras, and there are not enough particles in the Van Allen belts.

We now understand that the electrons that cause the auroras come in from the Earth's magnetic tail and are energized locally within the magnetosphere. The auroral oval is intermittently activated during sub-storms (see Sect. 8.4). Particle energization results from an increase in the circulation rate within the magnetosphere, primarily by the creation of electric fields that transport particles from the Earth's back-door storage region (the plasma sheet), thereby carrying enough currents of the type proposed by Birkeland. Such events occur typically a few times per day, but happen more often during active periods on the Sun when coronal mass ejections bring solar magnetic fields conducive to merging with Earth's field. Thus, it is really the Sun that controls the intensity of the auroras, like the dimming switch of a light.

When a solar eruption, with its associated shocks and magnetic fields, reaches the magnetosphere, the auroral oval intensifies and swells. Rare solar eruptions, with strong magnetic fields and high speeds, cause the auroral ovals to expand and move nearly as far south as the equator. For instance, about 18 hours after the white-light flare recorded by Carrington in 1859 (see Sect. 7.2), auroras were noticed as far south as Honolulu, Hawaii in the northern hemisphere and as far north as Santiago, Chile in the southern hemisphere. Such events, called great auroras, are seen in both hemispheres at the same time, but might only occur once or twice in a century. They are accompanied by world-wide disturbances of the Earth's magnetic field that we next discuss.

8.8 GEOMAGNETIC STORMS

A series of discoveries in the mid-nineteenth century led to the conclusion that solar activity causes the terrestrial magnetic field to vary on a global scale. In 1838 the German mathematician Karl Friedrich Gauss showed that our planet's magnetic field originates deep inside the Earth's core and extends far out into space with a dipolar shape. This suggested that world-wide magnetic observations would be required to understand terrestrial magnetism. Just a few years later, Heinrich Schwabe discovered the decade-long sunspot cycle (see Sect. 5.3). In the meantime, Great Britain had instigated a network of magnetic observations throughout its colonial empire, at least in part because compasses were used for ocean navigation. By 1852 Colonel Edward Sabine, superintendent of four of the magnetic observatories in the British colonies, was able to show that global magnetic fluctuations are synchronized with the sunspot cycle.

In response to a letter from the astronomer John Herschel, that called attention to Schwabe's sunspot results, Sabine wrote:

> With reference to Schwabe's period of 10 years having a minimum in 1843 and a maximum in 1848, it happens that by a most curious coincidence (if it be nothing more than a *coincidence*) that in a

> paper now waiting to be read at the Royal Society, I trace the very same years as those minimum and maximum of an apparent periodical inequality which took place in the frequency and magnitude of the [terrestrial] magnetic disturbances and in the magnitude of the mean monthly range of each of the 3 magnetic elements shown concurrently in the two hemispheres.[38]

Thus, the existence of global magnetic disturbances was established, and it was shown that the entire terrestrial magnetic field was subject to a systematic decade-long variation in tandem with the number of sunspots and associated magnetic activity on the Sun.

Occasionally compass needles fluctuate rapidly with a pronounced swing, indicating significant variations in the Earth's magnetic field. They are called geomagnetic storms. Unlike localized stormy weather on land or sea, a geomagnetic storm is invisible and silent, undetectable by the human eye or ear, but like our daily weather, geomagnetic storms can sometimes have devastating effects.

Shortly after the turn of the twentieth century, E. Walter Maunder, a sunspot expert from the Royal Observatory at Greenwich, England, found that the rare, great geomagnetic storms are associated with large sunspots near the center of the visible solar disk, and that many geomagnetic storms occur at intervals of 27 days, corresponding to the apparent rotation period of the Sun as viewed from the Earth. He concluded that this recurrence could be explained:

> by supposing that the Earth has encountered, time after time, a definite stream, a stream which, continually supplied from one and the same area of the Sun's surface, appears to us, at our distance, to be rotating with the same speed as the area from which it rises.[39]

If averaged over both a yearly and global scale, the geomagnetic storms vary in step with the sunspot cycle. When the Sun shows more spots, the terrestrial magnetic field is more frequently disturbed by vio-

FOCUS 8B
Changing Paradigms for Geomagnetic Storms

For at least half a century it has been supposed that the great non-recurrent geomagnetic storms, that shake the Earth's magnetic field to its very foundations, are caused by similarly infrequent but powerful solar flares detected by their intense radiation. It was thought that flare explosions blew clouds of magnetic fields and entrained material off the Sun that arrived at Earth one or two days later to cause severe disturbances of terrestrial magnetism.

Many solar astronomers now believe that this picture is incorrect. The Earth-bound clouds, or coronal mass ejections, are now thought to leave the Sun more or less of their own volition, and any associated flare is thought to be an effect rather than the cause of the mass ejection. It's as if the ejected coronal material, rather than being tossed out as an unwelcome house guest, celebrates its departure by sponsoring a local fireworks show. This analogy cannot be pressed too far because coronal mass ejections that cause great magnetic storms at Earth sometimes depart from the Sun rather unceremoniously in association with little attendant flare emission. But that's just what led some scientists to doubt that flares were the driving force behind major solar-terrestrial disturbances in the first place.

The new paradigm for great geomagnetic storms has been championed by Jack Gosling of the Los Alamos National Laboratory who showed that coronal mass ejections are well associated with the major storms, complementing other studies that indicated that fast interplanetary shocks and large energetic particle events were also closely associated with mass ejections. Not everyone accepts Gosling's view. In fact, the publication of his 1993 paper entitled "The Solar Flare Myth" in the *Journal of Geophysical Research* has touched off an ongoing, first-rate controversy that promises to keep the solar-terrestrial field astir for some time as both supporters and detractors probe questions raised by the new paradigm.

More recently, Nancy Crooker of Boston University and Ed Cliver at the Air Force's Phillips Laboratory have proposed yet another revisionist challenge to established views of the solar sources of geomagnetic storms – in this case, the recurrent variety. They argue that the coronal holes, thought for two decades to be the sources of the 27-day recurrent geomagnetic activity, are not the whole story. Instead, Crooker and Cliver go back to suggestions made early in the space age that the interaction of high-speed and slow-speed streams in the solar wind accounts for the most intense parts of the recurring storms.

In this picture, the long-mysterious source of recurrent geomagnetic storms, the M-regions, are no longer equated solely with coronal holes, the sources of high-speed streams, but also with their adjacent coronal streamers, the probable sources of the slow solar wind. They suggest that coronal mass ejections may additionally play a role in recurrent storms since the mass ejections often originate in the streamer belt. Like Gosling's research on non-recurrent storms, this study is likely to stimulate new controversy and investigations. That's just the way that science works!

lent storms. But it is not the sunspots themselves that bring about the changes on the Earth. The oscillating compass needle acts as a barometer of sporadic solar eruptions that occur more often when the Sun is more spotted and active.

As emphasized by Julius Bartels, recurrent geomagnetic storms sometimes occur when there are no visible sunspots. In 1932 Bartels referred to the sources of geomagnetic storms as M regions and noted that while M regions were sometimes associated with sunspots, in many other cases they seemed to have no visible counterpart. The M probably stood for magnetic, but it might have denoted mysterious; for several decades the elusive source of the recurrent geomagnetic storms remained one of the great unsolved mysteries in solar physics.

It was eventually found that the evasive M regions are invisible coronal holes observed in X-ray images of the Sun (see Sects. 6.3 and 6.5). Long-lived, high-speed streams in the solar wind, that emanate from the coronal holes, periodically sweep past the Earth, producing geomagnetic storms every 27 days. The recurrent storms turned out to be moderate or small in intensity, to commence gradually, and to dominate geomagnetic activity in the declining phase of the 11-year solar activity cycle.

Great, non-recurrent storms, that only occur sporadically at intervals of about a year and commence suddenly, are caused by interplanetary disturbances associated with coronal mass ejections (but see Focus 8B). Such outbursts occur most frequently when sunspots are most numerous, and may be energized by coronal magnetic fields that are linked to underlying sunspots (see Sect. 7.7). The radiation signatures of these intense solar eruptions are typically observed a day or two before the great, non-recurrent geomagnetic storms, and can be used to warn us of their impending consequences. (Magnetic storms, be they recurrent or non-recurrent, are always accompanied by intense substorms – see Sect. 8.4, and to a certain extent can be described as a series of intense substorms, although this view is debatable.)

The consequences of a great magnetic storm are truly awesome. Several times every solar cycle, a solar eruption of extraordinary energy creates a brief, violent gust in the solar wind that sets the entire magnetosphere reverberating with catastrophic impact. When the high-velocity shocks arrive at the Earth, followed by the magnetic fields, they can compress the dayside magnetosphere down to half its normal size. High rates of connection between the solar-wind magnetic field and the terrestrial one increase the size of the magnetotail that connects to the poles, and as a result the auroral oval intensifies and spreads eerily beautiful auroras across the sky to tropical latitudes in both hemispheres.

Auroral currents and electrons associated with the geomagnetic storm produce unpredictable, erratic heating and expansion of the upper atmosphere, causing satellite orbits to decay more quickly and less predictably than otherwise. (Atmospheric heating also gradually in-

creases as the result of enhanced radiation during the 11-year cycle of solar activity – see Sect. 8.9; similar or greater effects are caused by intense
geomagnetic storms that are less predictable and last a few days with
rapid onset.) Geostationary satellites that stay over the same spot on
Earth are also endangered when a great magnetic storm compresses the
magnetosphere from its usual location at about 10 Earth radii to below
their synchronous orbit at a distance of 6.6 Earth radii, exposing them to
the full, undeflected brunt of the solar wind.

The magnetism temporarily changes its size on a truly grand scale,
confusing geomagnetic navigational and detection sensors in satellites
and disorienting homing pigeons and other migratory animals that depend on the Earth's magnetic field for guidance. These great storms also
commonly disrupt radio communications and occasionally cause airplane pilots to lose contact with their control towers.

The disturbed magnetic fields induce electric currents in the Earth's
surface which can produce voltage surges on long-distance power lines
grounded at two widely-separated points. The surge can cause transformers at power plants to melt or be knocked out of service, overload-
ing electrical grid systems and sometimes darkening entire cities. During a particularly severe geomagnetic storm, on March 13, 1989, virtually
all of the Canadian province of Quebec was plunged into complete darkness without warning and within a few seconds. Six million customers
were left without electricity for over nine hours, costing around 500 million dollars counting losses only from unserved demand. (At the same
time that the networked power grid collapsed, large transformers failed
elsewhere in Canada and the United States.) As utility companies rely
more and more on long transmission lines to supply scattered demand
centers, they become increasingly vulnerable to these effects.

So, there is considerable practical interest in understanding and predicting the occurrence of great geomagnetic storms. The intense radiation and energetic particles from powerful solar eruptions create hazards to unprotected astronauts and also threaten satellites. We therefore
next consider these aspects of solar activity.

8.9 ASTRONAUTS AND SATELLITES IN DANGER

We are children of the space age, increasingly dependent upon
Earth-orbiting satellites and high-flying spacecraft. Geostationary
satellites, that hover above the same place on the Earth, relay and beam
down signals that are used for aviation and marine navigation, money
and commodity exchanges, and world-wide telephone communication.
Other satellites move in lower orbits and whip around the planet, scan-
ning air, land and sea for environmental change, weather forecasting and
military reconnaissance.

There is also a growing human presence in space. Airplanes can now take us just about anywhere in the world within a day or two. Astronauts plan to construct a giant space station in Earth orbit, return to the Moon and even visit Mars within the next few decades, perhaps as the first steps toward future extraterrestrial colonization.

The spacecraft of our informationally networked and space-faring society are increasingly at the mercy of the Sun, as they move within the charged, hostile environment of space. The solar wind continually feeds the space near Earth with particles, and eruptive events on the Sun can unpredictably release vast quantities of matter and energy, flooding the solar system with fast particles, magnetic fields and invisible radiation from radio waves to X-rays.

Eight minutes after an energetic solar flare, a strong blast of X-rays and ultraviolet radiation reaches the Earth and radically alters the structure of the planet's upper atmosphere; this in turn plays havoc with global communication systems. The radiation breaks apart individual atoms in the air, disrupting and transforming the ionized layer of the outer atmosphere. (This ionosphere is discussed in greater detail in Sect. 9.4.) During moderately intense flares, short wavelength, or high frequency, radio communications can be silenced over the Earth's entire sunlit hemisphere, and they do not return to normal until the flare disappears. (The night side is shielded by the Earth and remains unaffected.)

The United States Air Force and Navy are particularly concerned about this solar threat to radio communications. A solar flare once blacked out contact with a jet carrying President Ronald Reagan to China; for several hours the country's military Commander in Chief was unable to send or receive messages. More recently, in San Francisco, automatic garage doors began to open and close mysteriously, and burglar alarms went off for no apparent reason; these phenomena were eventually traced to a Navy backup radio system operating during a solar flare. The Air Force operates a global system of ground-based radio and optical telescopes and taps into the output of national, space-borne X-ray telescopes and particle detectors in order to continuously monitor the Sun for intense flares that might severely disrupt military communications and satellite surveillance.

Ultraviolet and X-ray emissions during periods of high solar flare activity can cause the temperatures in the upper atmosphere to soar to almost three times the values encountered at periods of low activity. A result of this heating is an outward expansion of the atmosphere and a concomitant increase in the gas density where satellites orbit. The greater atmospheric density leads to increased drag exerted on the satellites, pulling them into a lower orbit and sometimes causing ground controllers to temporarily lose contact with them.

Unlike the sudden increase in satellite drag associated with geomagnetic storms, that produce unpredictable, episodic effects, the flares create a long-term orbital degradation that is greatest at times of enhanced solar activity, mainly during the maximum of the 11-year sunspot cycle. The greater satellite drag at that time has sent some satellites spinning and tumbling out of control, and caused the rapid orbital decay of others, shortening their lifetime. Rising solar activity, for example, sent both the Skylab Space Station and the Solar Maximum Mission (SMM) satellite into an uncontrollable and fatal spiral toward the Earth; both spacecraft were ungratefully destroyed by the very phenomenon they were designed to study – solar flares. Scientists can nevertheless estimate satellite lifetimes by predicting the frequency and intensity of solar flares together with the associated changes in air density.

Energetic charged particles trapped in the Van Allen radiation belts provide a persistent, ever-present threat to satellites passing through them. The high-speed, subatomic particles can move right through the thin metallic skin of a spacecraft, damaging the microchip electronics inside. (Miniaturized electronic components have low mass and require little power, so they greatly increase a satellite's capability, but they are also more susceptible to radiation damage than their older, heavier and larger counterparts.) Metal shielding and radiation-hardened computer chips are used to guard against this recurrent hazard, and satellite orbits can be designed to minimize time in the radiation belts, or to avoid them altogether.

Nothing can be done to shield the solar cells used to power nearly all Earth-orbiting satellites; the photovoltaic cells convert sunlight to electricity and therefore have to be exposed to space. The danger was first realized in 1962 when an atmospheric nuclear explosion called *Starfish* produced artificial radiation belts, wiping out the solar arrays of several satellites in low-altitude orbits at the time. As satellites repeatedly pass through the Earth's natural radiation belts, they slowly deteriorate and shorten their useful lives, like parents that are continuously exposed to noisy children. (When the Van Allen radiation belts are inflated by exceptional solar activity, solar cells can age years in just a few days; however, the damaging high-speed magnetospheric electrons tend to increase at solar activity minimum, for as yet unknown reasons.)

The recurrent threat of moving within the radiation belts is particularly acute for satellites in low orbits that pass over an electronic danger zone, centered off the Brazilian coast. This region, known as the South Atlantic Anomaly, has been likened to the Sargasso Sea, a persistent but poorly known hazard to marine navigation. The South Atlantic Anomaly is caused by a displacement of the Earth's magnetic core by about 500 kilometers from the planet's center. As a result, particles trapped in the inner radiation belt can approach closer to the Earth's surface above the South Atlantic than elsewhere, so this region is anomalous because of its proximity. It can be thought of as a localized weakening or hernia of the Earth's protective magnetosphere. (The Earth's equatorial radius is 6378 kilometers, and the inner belt passes within a few hundred kilometers of its surface.)

When a low-flying spacecraft goes through the South Atlantic Anomaly, energetic charged particles can penetrate inside it and disrupt its computers or other scientific instruments. Scientists attempt to shield their instruments against the pervasive danger, and try not to use them when passing through it.

Infrequent, anomalously large eruptions on the Sun can hurl very energetic protons toward the Earth and elsewhere in space. When striking a spacecraft, they can wipe it out in a single devastating blow, sort of like Jack the giant killer. Such a solar energetic particle event has a tremendous energy (larger than 10 MeV) owing to the proton's relatively large mass and speed (approaching the velocity of light). It can cripple spacecraft microelectronics or solar cells through a single event upset. (In deep space, heavy cosmic-ray ions can also produce a single-event upset, with greater threat at solar minimum.)

On average, one such solar energetic particle event hits the Earth every month during the maximum of the 11-year cycle of solar activity, and perhaps once a year at the minimum. These averages are somewhat misleading, however, because solar events tend to cluster in time. Several of them might occur in one month followed by an interlude of a few months during which the Sun seemingly regenerates strength for a new assault. Such outbursts have already knocked several communication and weather satellites out of commission. Space weapons can produce a similar effect; so if you didn't know the Sun was at fault, you might think someone was trying to shoot down our satellites. In fact, the military monitors about 30 000 objects in space, keeping track of just where every one of them is located at any time of the day or night.

The solar protons are so energetic that there is little protection from the terrestrial magnetic field, at least for high-altitude geostationary satellites. And since the protons are moving so fast, there isn't much warning. The most energetic particles can arrive at the Earth within minutes after the associated outburst of radiation.

There is an ongoing controversy over their origin. Some say that the protons are accelerated to very high velocities and energies by powerful solar flares; others argue that they are propelled by the shocks that precede coronal mass ejections. Flares originate in more compact regions on the Sun, and their charged output will be guided along the curved magnetic highways of interplanetary space. If their protons are sent on a collision course with the Earth, the solar flare must occur on the Sun's western edge. In contrast, coronal mass ejections accelerate energetic particles over a broader region in space and are more likely to connect with the Earth. Both types of solar eruptions probably contribute to the solar proton flux at Earth, and since the controversy over the dominant process is not yet settled, we will not distinguish between their exact causes.

The stream of high-energy particles created during large solar eruptions causes concern for crews and passengers on commercial airplanes, especially those flying along polar routes. Airlines now have more planes operating at higher altitudes than ever before, above most of the protective blanket of the atmosphere, and over the North magnetic pole, where the Earth's magnetic field does not block the incoming solar particles. The particles are channeled along the magnetic field and penetrate to low altitudes in the polar regions, exposing passengers to elevated levels of radiation. The higher the plane is flying and the closer to the poles, the higher the radiation dose.

The health risk is highest for frequent fliers, pilots and flight attendants who travel polar routes often. The fatal cancer rate from in-flight radiation for a 20-year veteran is about 1 percent, far larger than that of second-hand smoke but less than the cancer risk faced by the average cigarette smoker. Pregnant women are definitely advised to avoid such flights because of risk to their unborn child.

There are even greater hazards to astronauts or military pilots who enter space at higher altitudes. According to an expert at the United States Defense Nuclear Agency, high-flying people can be provided with drugs that will make them temporarily survive a lethal dose of solar particles or radiation. It's a matter of patriotism and cost. They are going to die anyway, so why not take the injection, save the spacecraft, and come on home to die. There is no sense in being the first corpse in space.

Because of the potential genetic damage, astronauts are supposed to have had all their children before flying in outer space; otherwise they might have some very weird offspring. Probably because of hormones, men are more radiation-resistant than women, and the resistance peaks between the ages of forty five and fifty. So, if genetic harm and other health risks are the dominant factor, most astronauts will be middle-aged men, and they usually are.

Space is radioactive, and it's no laughing matter! Even in near-equatorial, low-altitude orbit, beneath the protection of Earth's magnetic field, astronauts have been troubled by energetic protons causing flashes in their eyes. (The energized particles rip through the satellite walls and pass through eyeballs, making them glow inside.)

Solar energetic particle events can endanger the health and even the lives of astronauts when they are unprotected by the Earth's magnetic field; that is, when they are in high inclination orbits or outside the Earth's magnetosphere on lunar or interplanetary flights. The shielding of a typical spacecraft is not then enough to protect a human from cataracts or skin cancer during a major solar energetic particle event, and the solar particles would be lethal for unprotected astronauts that venture into space (Fig. 8.16) to unload spacecraft cargo, construct a space station or walk on the Moon or Mars.

Fig. 8.16. Man in Space. Astronaut Donald Peterson, on a 50-foot tether line during his 4-hour, 3-orbit space walk, moving toward the tail of the Space Shuttle Challenger as it glides around the Earth. Hundreds of miles above the Earth, there is no air and an astronaut must wear a space-suit. It supplies the oxygen he needs and insulates his body from extreme heat and cold. However, a space-suit cannot protect an astronaut from energetic particles hurled out from the Sun; he must then be within the protective shielding of a spacecraft or experience possibly lethal radiation. (Courtesy of NASA)

A disaster has so far been avoided because previous stays on the Moon were of short duration (a few days) when no major solar eruption occurred. Luckily, a large solar flare took place in August 1972 between manned flights to the Moon in April and December 1972 (Apollo 16 and 17); it might have been lethal to the Apollo astronauts aboard their relatively unshielded vehicle.

A longer, future trip to Mars will involve considerable risks. The one-way trip could take about 9 months, and the entire voyage might last two or three years. Some estimate that every third human cell would be damaged during the flight. Of course, the frequency of lethal solar eruptions is rare (a few times a year) during the maximum in solar activity, and even less likely at the minimum. However, since there are more hazardous cosmic rays at solar minimum (see Focus 8A), a flight at that time would not be safe either.

Given adequate warning, future astronauts might seek temporary protection within thick-walled shelters in spacecraft or underground shelters on the Moon. Astronomers therefore use telescopes on the ground and *in situ* particle detectors or remote-sensing telescopes on satellites to carefully monitor solar activity. This may enable them to predict the short, violent storms and long, repetitive seasons of space weather, eventually permitting our fellow humans to use space with less risk. Accurate, specific forecasts of the great space weather disturbances, such as geomagnetic storms, intense solar radiation, and high-speed solar protons, should also enable evasive action that can reduce disruption or damage to communications, defense and weather satellites, as well as electrical power systems on the ground.

In the meantime, space physicists continue measurements that may decipher the complex, ever-changing, interacting regions of space near Earth.

8.10 CONTEMPORARY SOLAR-TERRESTRIAL RESEARCH

Solar-terrestrial research, in its broadest sense, investigates the dynamic interplay between the Sun and the Earth, using deep-space probes both within the Earth's magnetosphere and further out in the solar wind. These tools have given us the eyes to see the invisible and hands to touch what cannot be felt. Previous measurements in space have nevertheless been obtained piecemeal from observations taken at different times and locations. For the most part, the spacecraft were isolated from, and insensitive to, other critical areas of the space near Earth, or geospace.

We are only beginning to understand the interconnected processes that join the key regions of space into a working whole. Their boundaries are in constant motion, and changes in one region invariably stimulate subsequent changes or adjustments in others. They all act and depend on each other in a complex, interacting web that ultimately depends on the vagaries of the Sun. Individual, localized observations therefore yield a blurred image of our highly dynamic and changing environment. Attempts to understand ever-changing space from this incomplete picture is something like trying to assemble a puzzle with only a few pieces, and moreover those pieces vary in shape and size with time.

Now, in a new approach, space agencies from the United States, Japan, Russia, and the western European nations will combine their experience and resources to make coordinated, simultaneous, multi-spacecraft measurements of the Sun, the solar wind, and key areas of the Earth's magnetosphere (Fig. 8.17). For the first time, scientists will be aware of what is happening simultaneously in the major regions of the complicated, interacting system that links the Earth to its fundamental driving force – the Sun. It will provide a new global perspective on the intricate coupling between the Sun and Earth and help to understand the large-scale configuration and interplay of various processes in the magnetosphere and the enveloping solar wind.

Known as the International Solar-Terrestrial Physics (ISTP) Program, this effort will provide a new three-dimensional perspective of geospace. Each satellite of this international armada is targeted to a key region either inside or outside the magnetosphere (also see Fig. 8.17):

Fig. 8.17. International Solar-Terrestrial Research. A flotilla of international spacecraft will study solar-terrestrial interactions during the 1990s. The name of each spacecraft designates the region of the Earth's magnetosphere that will be investigated. Japan's Geotail spacecraft, launched by the United States in 1992, has a long, looping orbit that takes it through the magnetotail, one of the key regions where the solar wind enters the Earth's magnetosphere. NASA's Wind spacecraft, launched in 1994, studies the solar wind just before it impacts Earth's magnetosphere; while NASA's Polar spacecraft, to be launched in 1995, orbits the polar regions of the magnetosphere and observes the transfer of solar particles into the Earth's upper atmosphere where they cause the auroras. The Solar and Heliospheric Observatory (SOHO), a joint European Space Agency (ESA)-NASA mission to be launched in 1995 or 1996, will be placed at a libration point where the gravitational pull of the Sun and Earth are equal; SOHO will therefore be able to continuously study both the Sun and the solar wind in a region that has no night. Cluster, another joint ESA-NASA mission, consists of four closely positioned, identical spacecraft to be launched in 1995 to study the fine-scale structures of the magnetosphere and the solar wind – also see Fig. 8.8. (Courtesy of NASA)

1 Geotail Spacecraft
2 Interball Tail Spacecraft
3 Polar Spacecraft
4 Interball Auroral Spacecraft
5 Wind Spacecraft
6 Cluster Spacecraft
7 SOHO Spacecraft

Wind, where the solar wind first meets the Earth's environment; Polar, where solar particles at the polar cusps can penetrate into the atmosphere; and Geotail, where responses to the impact of the solar wind are felt deep in the magnetotail, over a million kilometers from Earth. Cluster's four closely positioned, identical spacecraft will conduct simultaneous measurements in the magnetosphere and the solar wind, while the Solar and Heliospheric Observatory (SOHO) will look back at the ultimate source of it all, the Sun, in order to identify and analyze the solar features that cause terrestrial effects, while also investigating the Sun's hidden interior (also see Focus 6B and Sect. 4.5).

The combined set of measurements, taken at a multitude of points in space, will not only specify how the variable Sun feeds energy into the space near Earth; but also how energy is coupled back and forth between the incident solar wind and the magnetosphere on the one hand and between the magnetosphere and the Earth's upper atmosphere on the other. We can then determine exactly how the active magnetosphere is driven by the solar wind, and also how it is internally sustained. Altogether, the spacecraft will provide a magnificent, global perspective of the Sun's effect on Earth – a major *tour de force* of the space age.

London, The Houses of Parliament: Sun Breaking through the Fog. 1904. Claude Monet wrote that he dreamed of painting the Sun "setting in an enormous ball of fire behind the Parliament", but the Sun's movement and changing appearance in the dreary London weather made this difficult. Instead, Monet used diffuse orange and reddish hues to capture the effect of sunset in a dense, foggy atmosphere, flattening the massive, neo-Gothic building to a dark silhouette of towers and pinnacles. (Courtesy of the Musée d'Orsay, Paris. Photograph: Musées Nationaux, Paris)

CHAPTER 9

Transforming
the Earth's Life-Sustaining Atmosphere

9.1 FRAGILE PLANET EARTH – THE VIEW FROM SPACE

More than two decades ago astronauts bound for the Moon looked back at our planet, seeing it suspended there all alone, swinging through the chill of outer space. For the first time we saw our home as a glistening blue and turquoise ball, light and round and shimmering like a bubble, flecked with delicate white clouds (Figs. 9.1 and 9.2). This perspective created a world-wide awareness of the Earth as a unique and vulnerable place, a tiny, fragile oasis in space.

The astronomer Fred Hoyle anticipated such an impact, declaring in 1948 that:

> Once a photograph of the Earth, taken from outside, is available, we shall, in an emotional sense, acquire an additional dimension ... .
> Once let the sheer isolation of the Earth become plain to every man whatever his nationality or creed, and a new idea as powerful as any in history will be let loose.[40]

To our ancestors only a few centuries ago, the ocean, land and sky seemed vast and almost limitless. But all that has changed. Today, the entire world is our playground.

And it is only now that we can see it from space, that we realize the magnitude of what we are doing to the Earth. Indeed, it is probably no accident that the advent of the space age has coincided with a growing concern about the terrestrial environment and the precarious fate of life within it.

Only from space can we see the living planet Earth as a whole, as a single, unified system encompassing life on land and in the sea (Fig. 9.3). Spacecraft monitor its vital signs; weather satellites are an example. Powerful instruments on satellites can also zero in to take a magnified birds-eye view of the Earth's ever-changing surface (Fig. 9.4).

Astronauts can see the horizon as a curved line, showing a thin membrane of air that ventilates, protects and incubates us (Fig. 9.5). As astronaut Jim Buchli described it in the *Blue Planet*:

Fig. 9.1. The Water Planet. Almost three-quarters of the Earth's surface is covered with water, as seen in this view of the North Pacific Ocean. Earth is the only planet in the solar system where substantial amounts of water exist in all three possible forms – gas (water vapor), liquid and solid (ice). Here white clouds of water ice swirl just below Alaska; the predominantly white ground area, consisting of snow and ice, is the Kamchatka Peninsula of Siberia. Japan appears near the horizon. From this orientation in space, we also see both the day and night sides of our home planet. (Courtesy of NASA)

Fig. 9.2. Dawn. The morning Sun reflects from the Gulf of Mexico and the Atlantic Ocean, as seen from the Apollo 7 spacecraft. Most of the Florida peninsula appears as a dark silhouette. (Courtesy of NASA)

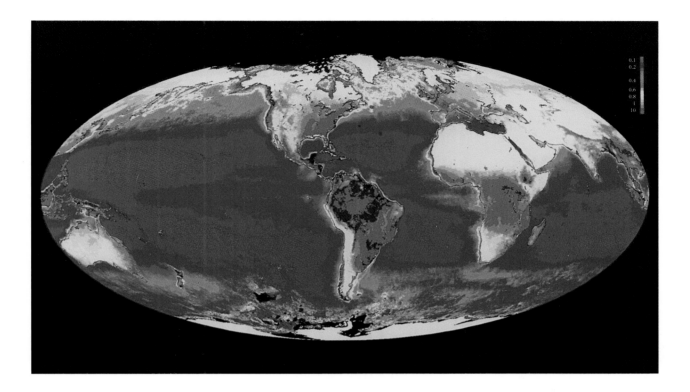

Fig. 9.3. Global Biosphere. The first truly global view of the Earth's biosphere, showing life on land and sea. The ocean portion shows the concentration of plankton, microscopic plants that grow in the upper sunlit portions of the ocean and are the ultimate source of food for most marine life. At sea, the red and orange colors denote the highest concentration of plankton; yellow and green represent areas of moderate plankton concentration, and blue and violet describe the lowest concentrations. The land vegetation image shows rain forests (*dark green areas*), tropical and subtropical forests (*light green*) and areas of low vegetation and deserts (*yellow*) – color scale on the upper right side. This image combines years of data from two satellites, Nimbus 7 – sea and NOAA 7 – land. (Courtesy of G. C. Feldman and C. J. Tucker, NASA)

Look at how thin the atmosphere is … . Everything beyond that thin blue line is the void of space. And everything below it is what it takes to sustain life. And everything that we do to this environment, and our quality of life, is below that little thin blue line. That's the only difference between what we enjoy here on Earth and the really harsh uninhabitable blackness of space. It's not very wide, is it?[41]

The Earth's diameter is one thousand times greater than the width of the atmosphere surrounding it. Indeed, the distance from the ground to the top of the sky is only about 10 kilometers, or no farther than you might run in an hour.

The tolerances of life are so narrow, that small variations in the flow of the Sun's energy can spell the difference between a benign and hostile environment for life on Earth. The Earth's protective ozone layer can be enhanced or depleted by changes in the Sun's radiation (see Sect. 9.6); fluctuations in solar radiation might account for global temperature changes observed during the past century (see Sect. 10.3); and changes in the amount and distribution of sunlight on the Earth produce both the ice ages and the warm interglacial periods (see Sect. 10.6). Indeed minor changes in the composition of our air, or seemingly insignificant variations in the amount of solar radiation reaching it, could transform our globe into an unlivable place.

Fig. 9.4. Mojave Desert. This color-enhanced Landsat image, obtained primarily at infrared wavelengths, reveals previously undetected faults cutting through the central and eastern Mojave Desert. The vivid colors show alignments of contrasted rock types. (Courtesy of Jurrie Van Der Woude, Jet Propulsion Laboratory)

Fig. 9.5. A Thin Colored Line. A brilliant red marks the thin atmosphere that warms and protects us, as viewed from space at sunrise over the Pacific Ocean. Without this thin atmospheric membrane we could not breathe and water would freeze. The overlying ionosphere, colored blue, is created by the Sun's energetic X-ray radiation. (Courtesy of NASA)   ▽

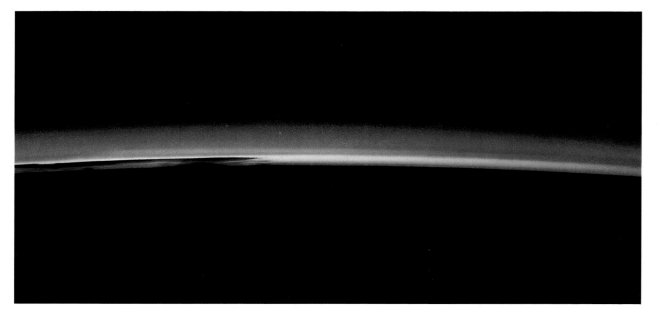

9.2 THE LIFE-SAVING GREENHOUSE EFFECT

Our planet's surface is now comfortably warm because the atmosphere traps some of the Sun's heat and keeps it near the surface. The thin blanket of gas acts like a one-way filter, allowing sunlight through to warm the surface but preventing the escape of some of the heat into the cold, unfillable sink of space. Much of the ground's heat is reradiated out toward space in the form of longer infrared waves that are less energetic than visible ones and thus do not slice through the atmosphere's gas as easily as sunlight.

These gases absorb heat at infrared wavelengths, retaining it and raising the temperature of the planet. The atmosphere's warming influence has been dubbed the "natural" greenhouse effect to distinguish it from global greenhouse warming caused by "unnatural" human activity.

Right now, this "natural" greenhouse is literally a matter of life and death. It keeps the planet 31 degrees Celsius hotter than it would otherwise be. Without this greenhouse warming, the oceans would freeze and life as we know it would not exist. The average temperature of the Earth's surface is now about 15 degrees Celsius; so without the greenhouse effect it would be well below the freezing point of water and comparable to the average temperature of the airless Moon.

The idea that our atmospheric blanket might warm the Earth is not new. It was apparently suggested in 1827 by the French mathematician Jean-Baptiste Fourier and developed by the British scientist John Tyndall in the 1860s. (Fourier is now well known for his investigations of waves and other periodic motion, a branch of mathematics now known as Fourier analysis, and Tyndall explained why the sky is blue.) Fourier maintained that the atmosphere acts like the glass of a hothouse, or greenhouse, because it lets through the visible light rays of the Sun but retains invisible heat rays from the ground.

Tyndall measured the transmission of infrared heat radiation through different gases, and was astonished to find that significant heat is absorbed by minor ingredients of the air, like water vapor and carbon dioxide. As he expressed it in 1861:

> This aqueous vapour, which exercises such a destructive action on the [Sun's] obscure [heat] rays, is comparatively transparent to the rays of light. Hence the differential action, as regards the heat coming from the Sun to the Earth and that radiated from the Earth into space, is vastly augmented by the aqueous vapor of the atmosphere .... Similar remarks would apply to the carbonic acid [carbon dioxide] diffused through the air.[42]

As Tyndall also showed, the main constituents of the atmosphere, nitrogen (78 percent) and oxygen (21 percent) play no part in the greenhouse effect. These diatomic molecules are incapable of absorbing infrared radiation. In contrast, water vapor and carbon dioxide molecules consist of three atoms that are less constrained, so they absorb the heat radiation.

The Swedish chemist, Svante Arrhenius, provided strong quantitative support for these ideas by determining the amount of heat absorbed by carbon dioxide and water vapor in the atmosphere. He concluded that minor variations in the amounts of these substances could produce either the ice ages or the warmest intervals between them. In his own words, written in 1896:

> A simple calculation shows that the [ground] temperature in the
> Arctic regions would rise 8 to 9 degrees Celsius if the carbonic acid
> [carbon dioxide] increased to 2.5 or 3 times its present value.
> In order to get the temperature of the ice age, the carbonic acid in
> the air should sink to 0.65 to 0.55 of its present value, lowering
> the temperature by 4 to 5 degrees Celsius.[43]

We now believe that changes in the global distribution of sunlight bring on the ice ages (see Sect. 10.6). Nevertheless, Arrhenius made a good estimate of the global warming effect of doubling the amount of carbon dioxide in our air, without the use of modern calculators or extensive computer models. He also pointed out that industrial activity was then noticeably increasing the amount of atmospheric carbon dioxide.

Why doesn't the atmosphere just keep heating up until it explodes? Provided that the Sun's radiative output stays constant, and that the composition of the atmosphere does not change, the greenhouse warming rises to a fixed temperature that balances the heat input from sunlight and the heat radiated into space. The level of water in a pond similarly remains much the same even though water is running in one end and out the other.

With greater energy from the Sun, or with more water vapor or carbon dioxide in the air, the balance point will shift to a higher average global temperature. This is why it becomes hotter in the summer when more sunlight directly illuminates the Earth's surface. It's also why cloudy nights tend to be warmer than clear nights; escaping infrared heat radiation is blocked by the water vapor in the clouds, keeping the ground warm at night.

Of course, you can have too much of a good thing, like lying in the Sun all day in the summer or eating too much ice cream too fast. The surface of Venus, under a dense atmosphere of carbon dioxide, has become hot enough to melt lead. The high-temperature world has been boiled dry, like a kettle left on the stove. Nothing could live there. Mars has only a feeble atmosphere with a meager greenhouse effect. This frozen world is locked in a permanent ice age.

9.3 THE EARTH'S CHANGING ATMOSPHERE

Nothing in the cosmos is fixed and unchanging, everything moves and evolves. People live out their changing lives; stars are born and die; the entire Universe is expanding from its explosive birth long ago. Sometimes the changes are slow and gradual, often they are quick and violent. In fact, the only unchanging thing in the Universe is change itself, and both the Sun's radiative output and the Earth's atmosphere are no exception.

It is thought that the Earth and other planets formed, together with the Sun, about 4.6 billion years ago. As the body of the new-born Earth accumulated and warmed up inside, the early atmosphere emerged from ancient volcanoes that spewed lava and released gases that were once trapped in the rock. The ancient oceans were also steamed out of the Earth's interior by the erupting volcanoes, condensing from water vapor in the primitive atmosphere. (Cosmic debris, attracted by the gravitational forces of the young Earth, probably slammed into its surface, vaporizing rocks, liberating gas and further enhancing the primitive air.)

The Sun began its life shining with only 70 percent of its present luminosity, slowly growing in luminous intensity as it aged. This steady, inexorable increase in the Sun's brightness is a consequence of increasing amounts of helium in the Sun's core; the greater mean density produces higher core temperatures, faster nuclear reactions, and a steady increase in luminosity.

If the Sun was much dimmer eons ago, the Earth's oceans would have been frozen solid even if the present greenhouse effect operated back then. Yet, sedimentary rocks, which must have been deposited in liquid water, date from 3.8 billion years ago, or less than a billion years after the Earth was formed. Moreover, there is fossil evidence for life about 3.5 billion years ago, and living things seem to have thrived in a warm, liquid environment ever since. For billions of years the Earth's surface temperature was not very different from today, even though the Sun's brightness has increased by some 30 percent since its creation.

The discrepancy between the warm climate record and the predictions of an early cold climate based on the Sun's faint luminosity has come to be known as the faint-young-Sun paradox. The paradox can be resolved if the Earth's primitive atmosphere contained about a thousand times more carbon dioxide than it does now. The greater heating of the enhanced greenhouse effect could have kept the oceans from freezing, ensuring that our planet stayed warm when it was young and protecting its surface from the chilling effect of a faint Sun.

As time goes on, the Sun keeps on getting brighter and brighter, and hotter and hotter. The Earth could only maintain a temperate climate so far by turning down its greenhouse effect as the Sun has turned up the heat. Otherwise the oceans would have boiled away by now and the Sun's heat would have wiped out life. So, the Earth's atmosphere, rocks and ocean apparently combine to provide a global thermostat, decreasing the amount of carbon dioxide in the air as the Sun brightens.

When the Sun becomes more luminous, the Earth's surface temperature initially increases. But more water is then evaporated, causing greater rain and weathering, and carbon dioxide is then removed from the atmosphere, lessening the greenhouse effect and lowering the temperature. (This may be partially offset by additional heat trapping of the water vapor molecules.) The process has been so effective that carbon dioxide now accounts for only 0.036 percent of the Earth's atmosphere. (The carbon dioxide content of our air has been reduced to 0.028 percent by natural processes, but it has now increased to 0.036 percent by industrial emission and deforestation.)

The reduction occurs because carbon dioxide from the atmosphere dissolves in rainwater, forming carbonic acid that eats away at rocks and releases elements that eventually run into the sea where plankton and shellfish use them to make calcium carbonate shells. They eventually die, falling to the ocean floor and building up layers of carbonate-rich sediment. The sedimentary deposits are now so extensive that there is currently as much carbon dioxide stored in sedimentary rocks on Earth, such as limestone and dolomite, as there now is in the atmosphere of Venus, where a runaway greenhouse effect has created temperatures hot enough to destroy life.

Plants and animals also play a role in transforming the atmosphere and determining its composition. Billions of years ago, when the young Earth's volcanic surface was too hot for living things, there must have been very little oxygen in the atmosphere. It was created by the early bloom of life and accumulated slowly. (Blue-green algae in the sea began to put oxygen in the air about 2 billion years ago; by about 400 million years ago, they had pumped sufficient oxygen into the atmosphere to sustain animals that breathe oxygen and live on land.) Free oxygen is now a major ingredient of the atmosphere, making up about a fifth of our air.

If plants did not continuously replenish the oxygen, it would eventually be depleted by animals and humanity that breathe oxygen. Even without animals, the oxygen gas would react chemically with other elements and be locked away in compounds like carbon dioxide and water or within rocks in about 4 million years. However, because ours is a living world, chemically reactive oxygen remains in the air.

Life may have developed the capability to control its environment, keeping it comfortable for living things in spite of threatening changes. This theory, developed by British inventor and scientist James Lovelock and American microbiologist Lynn Margulis, has been called the Gaia (pronounced GUY-ah) hypothesis, after the Greek goddess of mother Earth. (The name was suggested by Lovelock's neighbor, Nobel-Prize winner William Golding, author of the book *Lord of the Flies*).

According to this controversial concept, the physical world of rocks, oceans and atmosphere is affected by plant and animal life in such a way as to maintain the conditions that are conducive to life itself. After all, physical and chemical conditions on the Earth's surface have remained favorable for life during the past 3 billion years in spite of the Sun's slowly increasing brightness. As Lovelock and Margulis stated it in 1974:

> The present knowledge of the early environment suggests strongly that a first task of life was to secure the environment against adverse physical and chemical change. Such security could only come from the active process of homeostasis in which unfavourable tendencies could be sensed and counter measures operated before irreversible damage had been done.[44]

But there is probably no escape in the end. Whatever life on Earth does, its remote future is not secure! As the Sun brightens and turns on the heat, the greenhouse thermostat could remove all of the carbon dioxide from the atmosphere, starving the plants, and the animals that eat them, to death. (The increase in carbon dioxide by human burning of fossil fuels is apparently negligible compared to this long-term depletion.) And if life somehow manages to avoid this catastrophe, water vapor evaporating from the warmer oceans will produce an enhanced greenhouse effect that will warm the surface in tandem with the brightening Sun. Astronomers calculate that the Sun will certainly be hot enough in 3 billion years to boil the oceans away, leaving the planet a burned-out cinder, a dead and sterile place.

If that doesn't wipe us out, the expanding Sun will. It is expected to balloon into a giant star in about 7 billion years, engulfing the planet Mercury and becoming two thousand times brighter than it is now. That will be bright enough to melt the Earth's surface, and to turn the icy moons of the giant planets into globes of liquid water. The only imaginable escape would then be interplanetary migration to distant planets with a warm pleasant climate. (The swollen Sun will eventually expand out to the Earth's present orbit, but by this time the Sun will have shed considerable mass; because of the resultant weakening of the Sun's gravitational grip, the Earth will have moved slightly outside the enlarged Sun, keeping it from being swallowed by the Sun.)

So, our long-term prospects aren't all that great. Moreover, there are threats on a much shorter time interval that may be caused by either the ever-changing Sun or by human tinkering with the environment. To understand these effects, we must first consider how variable solar radiation transforms our air, heating it up and producing distinct layers within it.

9.4 OUR SUN-LAYERED ATMOSPHERE

On a clear day you can see forever, or at least for a long distance through the air. And on a warm, dry, windless day we are unaware of the touch of the air upon our skin. The drift of a cloud occasionally reminds us of the atmosphere about us, and on cold days we feel the air against our skin. The touch of the wind and the sight of birds and airplanes supported by their motion proves that there is something substantial surrounding us.

The Earth's gravity pulls down on the atmosphere, so the air near the ground is compressed and squashed by the weight of the overlying air. Like all gases, air is highly compressible, and because its molecules are mainly far apart, a gas is mostly empty space and can always be squeezed into a smaller volume. The air near the ground is compressed to greater density by the weight of the overlying air. Yet, even at the bottom of the atmosphere the density is only about a thousandth of that of liquid water; an entire liter of this air will weigh only one gram.

At greater heights there is less air pushing down from above, so the compression is less and the density of air gradually falls off into the vacuum of space. At a height of 10 kilometers (slightly higher than Mount Everest), the density of air has dropped to 10 percent of its value near the ground. No insects and few birds can fly in such rarefied air. At altitudes above about 50 kilometers, the air is too thin to support a jet airplane.

We can almost see the thinning of the air at greater heights when watching a group of hawks circling above a meadow or open field. As air near the warm ground heats it expands and becomes less dense, rising into the thinner atmosphere above. The rising air carries heat from the ground and distributes it at higher levels, giving free rides to the soaring birds. Hawks riding the currents of heated air sometimes rise so abruptly that it looks as if they were lifted and jerked up by strings. (Hot air also rises around the flame of a candle, and the inflowing air replenishes the supply of oxygen; without this current, the candle quickly goes out, as it would in a spaceship, where there is little gravity and no compression of the air.)

But why isn't the sky falling, as Chicken Little would have it, if gravity is relentlessly pulling it down? Because the air is warmed by the Sun, its molecules are in continuous motion and collide with each other, producing a pressure that prevents the atmosphere from collapsing to the ground.

At greater heights, the air is thinner so there are fewer molecular collisions and less air pressure. The decrease in air pressure with height accounts for the rise of balloons. When filled with a light gas, a balloon is buoyed upwards by the greater pressure of the air beneath it. If the upward force of buoyancy exactly matches the downward weight of the balloon and its contents, the balloon will remain suspended in the air.

Fig. 9.6. Layered Atmosphere.
The temperature of our atmosphere
tends to fall at higher altitudes where
the air expands in the lower pressure
and becomes cooler. But the temperature actually increases in two critical
regions that are heated by the Sun's
invisible radiation. They are the
stratosphere, with its critical ozone
layer, and the ionosphere. The stratosphere is mainly heated by ultraviolet
radiation from the Sun, while the
ionosphere is largely due to solar
X-ray radiation.

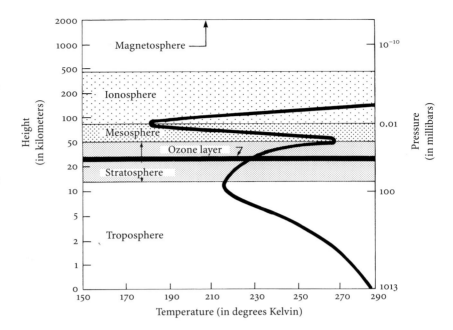

Not only does the atmospheric pressure decrease as we go upward,
the temperature of the air also changes, but it is not a simple fall-off. It
falls and rises in two full cycles as we move off into space (Fig. 9.6). Visible sunlight passes harmlessly through the atmosphere to warm the
ground and the air near it. The Sun also emits invisible ultraviolet and X-
ray radiation that heat the top portion of our atmosphere and arrange it
like a layer cake, each layer absorbing different wavelengths of the Sun's
radiation.

The lowest region, where all our weather occurs, is controlled by visible sunlight. Global atmospheric circulation, driven by differential solar
heating of the equatorial and polar surface, create complex, wheeling
patterns of weather in this region, leading to the designation troposphere, from the Greek *tropo* for turning.

The temperature decreases steadily with height in the ground-hug-
ging troposphere. When warm currents move up from the Earth's surface, the air expands in the lower pressure and becomes cooler. Most of
us have experienced the cold temperatures of mountain altitudes, and pilots of jetliners occasionally report bitter subzero temperatures outside
the plane. The average air temperature drops below the freezing point of
water (0 degrees Celsius or 273 degrees Kelvin) at only a kilometer or two
above the Earth's surface, and bottoms out at a height of about 12 kilometers above sea level. This is the greatest height achieved by the air currents, and it is the top of the troposphere. (The average extent of the troposphere varies with latitude, from about 16 kilometers above the warm
equator to roughly 8 kilometers over the cold poles.)

The top of the troposphere was detected near the end of the nineteenth century when Leon Philippe Teisserenc de Bort, a French meteorologist, launched hundreds of unmanned balloons that carried thermometers and barometers to altitudes as great as 15 kilometers. At this height, the temperature no longer decreased with altitude, and seemed to remain nearly constant. If the temperature was unchanging, the ingredients of the atmosphere above the troposphere might settle down into layers, or strata, depending on their weight, so de Bort named this region the stratosphere.

Contrary to everyone's expectations, the temperature increases at greater heights within the stratosphere, rising to nearly ground-level temperatures at about 50 kilometers above the Earth's surface; but we still use the name stratosphere to designate the layer of the atmosphere that lies immediately above the troposphere. (The temperature increases within the stratosphere because ozone molecules absorb the ultraviolet portion of the incoming solar energy – see Sect. 9.6).

The rise in temperature within the stratosphere was suggested by the booming artillery guns of World War I. They were heard at unexpectedly large distances, indicating that their sound waves were being bent around by the stratosphere and reflected or mirrored at an angle back down to the ground. Sound travels in much the same way for large distances across the cool surface of a lake in summertime, reflected by the warm air above it. (Waves of sound are similarly trapped in a relatively cool region sandwiched between two high-temperature ones in the outer parts of the Sun – see Sect. 4.1).

Above the stratosphere we come to the mesosphere, from the Greek *meso* for intermediate. The temperature declines rapidly with increasing height in the mesosphere, from cold temperatures at about 50 kilometers altitude to far below freezing at about 85 kilometers, where the temperature reaches the lowest levels in the entire atmosphere. The main reason for the decreasing temperatures is the falling ozone concentration and decreased absorption of solar ultraviolet rays.

The mesosphere has been known as the "ignorosphere" because it is too high to be reached by airplanes and too low to be studied by most spacecraft. The air at this height is too thin to support research balloons or aircraft, but thick enough for air drag to cause satellites to decay quickly from orbit. Sounding rockets pass through this region too rapidly to permit detailed study.

In December 1901, Guglielmo Marconi startled the world by transmitting a radio signal from England to Newfoundland. (The Italian scientist became an international hero, established the American Marconi Company, which later evolved into the Radio Corporation of America, or RCA, and in 1909 received the Nobel Prize in Physics for his unexpected long-distance communication.) Because radio waves travel in straight lines, and cannot pass through the solid Earth, no one expected that Marconi could send a radio signal halfway around the world.

Radio waves get around the Earth's curvature by reflection from an electrically charged layer, now called the ionosphere, extending into space from roughly 70 kilometers above the Earth's surface. (This explanation was provided by Arthur E. Kennelly, then at the Harvard School of Engineering, in early 1902, and almost simultaneously by Oliver Heaviside in England.)

The rapid expansion of radio broadcasting in the 1920s, as well as the concurrent development of pulsed radio signals, helped specify the structure and daily variation of the ionosphere. Edward Victor Appleton and his students measured the height of the reflecting layer by determining the elapsed time between transmitting a radio pulse and receiving its echo; like all electromagnetic radiation, the radio waves travel at the velocity of light. They showed that there are at least three such reflecting layers, now labeled D, E and F, at respective altitudes of 70, 100 and 200–300 kilometers. The three layers collectively account for much of what we today call the ionosphere. (In 1947 Sir Edward Appleton received the Nobel Prize in Physics for his investigations of the upper atmosphere, especially for the discovery of the so-called Appleton layers.)

Radio waves are not mirrored by the ionosphere unless their wavelength is longer than a certain critical value. (This critical wavelength provides a measure of the electron density in the ionosphere; the shorter it is the larger the density.) Radio waves with wavelengths that are less than the critical value can pass right through the ionosphere, because, roughly speaking, they are short enough to pass among the electrons. These shorter radio and microwave wavelengths are used in communications with satellites or other spacecraft in the space beyond the ionosphere.

The mystery of exactly what produces and controls the ionosphere was not solved until after World War II, when captured German V-2 rockets were brought to the United States. These and subsequent rockets, built by American engineers, were used by the Naval Research Laboratory to loft spectrographs above the atmosphere, showing that the Sun emits very energetic radiation at invisible X-ray and ultraviolet wavelengths. When this radiation reaches the upper atmosphere it rips electrons off the nitrogen and oxygen molecules, producing free electrons and atomic and molecular ions. The ionosphere above your head therefore develops as the Sun rises and decays as the Sun sets; it lingers on during the night but is not energized then.

Appleton's D, E, and F layers are explained by different wavelengths of the Sun's radiation that penetrate to varying depths in the atmosphere. Solar X-rays with the shortest wavelengths have the greatest energy and penetrate further before being completely absorbed, and the less-energetic rays with the longer wavelengths expend their ionizing capability higher up. Of course, they all have enough punch to knock electrons free of the strong forces that bind them within atoms. Most of the longer-wavelength solar ultraviolet radiation passes right through the ionosphere to the underlying stratosphere, but some of it at the shorter wavelengths, called the extreme ultraviolet, has enough energy to help create the ionosphere.

The temperature begins to rise again with altitude in the ionosphere. By absorbing the extreme ultraviolet and X-ray portion of the Sun's energy, the atmospheric molecules are broken into pieces, or dissociated, and their constituent atoms ionized; both processes release heat to warm the ionosphere. Here the temperatures skyrocket to higher values than anywhere else in the entire atmosphere. Indeed, some scientists prefer to call this region the thermosphere, or "hot" sphere. The thermosphere overlaps the E and F regions of the ionosphere, beginning at about 90 kilometers and extending upward to about 500 kilometers.

At higher altitudes, the atmosphere thins out into the exosphere, or the "outside" sphere. In the exosphere the gas density is so low that an atom can completely orbit the Earth without colliding with another atom. The temperature is so hot out there, and the atoms move so fast, that some atoms can escape the Earth's gravitational pull and travel into outer space. It is therefore the ionosphere and thermosphere that cap our Sun-layered atmosphere and provide the Earth's threshold into space.

Thus, it is the Sun's radiation that heats our atmosphere and creates layers within it. However, the Sun's steady warmth and brightness are illusory! The energy radiated from the Sun varies continuously at all wavelengths and on all time scales. These variations can dramatically alter our Sun-layered atmosphere and affect life beneath it.

9.5 THE INCONSTANT SUN

Day after day the Sun rises and sets in an endless cycle, an apparently unchanging ball of fire whose heat and light make life on Earth possible. The total amount of this life-sustaining energy has been called the "solar constant", because no variations could be detected in it, at least from beneath the Earth's changing atmosphere. (The solar constant is defined as the average amount of radiant solar energy per unit time per unit area reaching the top of our atmosphere at the Earth's mean distance from the Sun, amounting to about 1370 watts of power per square meter.) Indeed, only a decade or so ago, most astronomers and climatologists clung tenaciously to the view that the Sun shines steadily.

Yet, reliable as the Sun appears, it is an inconstant companion. The Sun actually fades and brightens in step with the Sun's changing activity level, and no portion of the Sun's radiative output is invariant. Indeed, the Sun's radiation is variable on all measurable time scales, and this inconstant behavior can be traced to the pervasive role played by magnetic fields in the solar atmosphere.

The recent discovery of variations in the total solar radiative output, or solar constant, was nevertheless somewhat unexpected. For instance, a group led by Charles Greeley Abbot, secretary of the Smithsonian Institution from 1919 to 1944, carried out daily measurements of the solar constant for decades, including expeditions to four continents and from sea level to mountain tops. Abbot concluded that "the Sun is a variable star" and suggested that changes in the solar constant were correlated with terrestrial weather patterns, but his colleagues concluded that if the Sun's total irradiance varied at all, it was at a level of less than 1.0 percent over decades. The fluctuations noticed by Abbot were attributed to atmospheric attenuation and distortions or improperly calibrated instruments.

The solar constant is actually remarkably constant, rarely changing by more than 0.2 percent, or one part in 500. Still, the Sun puts out so much energy that even this small variation has important implications. Unequivocal detection of such changes required the development of extremely stable and precise radiometers that could be lofted above the atmosphere in satellites. (Our air filters out invisible solar radiation and distorts the visible part; satellites provide a stable platform without these atmospheric limitations.)

Secure, definite evidence of the changing solar constant was not obtained until 1980 (Fig. 9.7). An exquisitely sensitive detector, created by Richard C. Willson of the Jet Propulsion Laboratory in Pasadena, California, measured the solar-constant value with an incredible precision of 0.01 percent or one part in 10 000, when averaged over one day. (His active cavity radiometer irradiance monitor, or ACRIM for short, was placed aboard the Solar Maximum Mission (SMM) satellite, launched in February 1980.) At this level of precision, the Sun's total radiative output is almost always changing, at amounts of up to a few tenths of a percent and on all time scales from 1 second to 10 years and possibly centuries.

Similar variations were also detected by the Earth Radiation Budget (ERB) radiometer aboard the Nimbus 7 satellite, launched in October 1978, with an even longer record, but with less precision. (The ERB was devel-

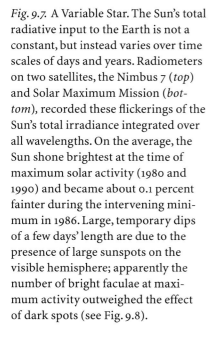

Fig. 9.7. A Variable Star. The Sun's total radiative input to the Earth is not a constant, but instead varies over time scales of days and years. Radiometers on two satellites, the Nimbus 7 (*top*) and Solar Maximum Mission (*bottom*), recorded these flickerings of the Sun's total irradiance integrated over all wavelengths. On the average, the Sun shone brightest at the time of maximum solar activity (1980 and 1990) and became about 0.1 percent fainter during the intervening minimum in 1986. Large, temporary dips of a few days' length are due to the presence of large sunspots on the visible hemisphere; apparently the number of bright faculae at maximum activity outweighed the effect of dark spots (see Fig. 9.8).

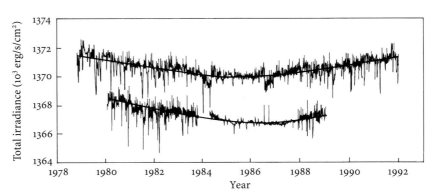

oped by John R. Hickey and his colleagues of the Eppley Laboratory in Newport, Rhode Island.) The two data sets are offset by roughly 0.2 percent in Fig. 9.7 due to a real disagreement in absolute calibration of the two radiometers, but the precision of each is much higher. Thus, when observed with sufficient exactness, the solar constant is not constant after all.

Hugh S. Hudson, a solar physicist at the University of California at San Diego, teamed up with Willson to help him identify the irradiance variations with well-known features on the Sun. They showed that the largest downward excursions or dips correspond to, and are explained by, the rotation of a large group of sunspots across the face of the Sun. The concentrated magnetism in sunspots apparently acts as a valve, blocking the energy outflow, which after all is why sunspots are dark and cool, producing reductions in the solar constant of up to 0.3 percent that last a few days.

During the first few years of their operation, the satellite-borne radiometers showed that the Sun's energy output was falling, as the number of sunspots decreased, slightly but steadily at the rate of about 0.1 percent in five years from 1980 to 1985. This fading was too fast to continue for long. If the Sun continued to dim at this rate, a terrestrial ice age would occur in several years. Fortunately, the observed decline bottomed out and the Sun brightened by a comparable amount during the next five years when the number of sunspots increased.

This meant that the varying solar output is geared to the 11-year sunspot cycle of magnetic activity. But surprisingly, the Sun becomes brighter and more luminous, rather than darker, as the number of sunspots on its surface becomes larger. This is not what you would expect from sunspots alone, for they dim the Sun's radiative output.

The increased luminosity is caused by bright patches on the visible solar surface. They are called *faculae*, from the Latin for little torches. The largest faculae are found in the vicinity of sunspots, but outside of these there is an enhanced network of comparatively bright although smaller structures, that cover the whole Sun. (The network brightenings are best seen in the chromosphere, where the brightest regions are called plages, while the name faculae is reserved for bright places in the photosphere.)

The decade-long variations in the Sun's total radiative output are the result of a competition between dark sunspots and bright faculae, with the faculae winning out. As magnetic activity increases, they both appear more often, but the excess radiance from bright faculae is greater than the loss from sunspots. This has been demonstrated by Peter V. Foukal, of Cambridge Research and Instrumentation, Inc., in Massachusetts, and Judith Lean of the Naval Research Laboratory in Washington, D. C. (Fig. 9.8). They showed that there is a good correlation between the long-term, solar constant variations and facular emission from the entire solar disk, once the effects of sunspot dimming have been removed.

Fig. 9.8. Facular Brightening. Most of the brightening of the active Sun can be attributed to photospheric magnetic structures called faculae, whose areas increase at times of high solar activity and sunspot number. Here the total irradiance, S, has been corrected for sunspot dimming, P; the residuals, S–P$_S$ (*dark line*) are closely correlated with facular area (*light curve*). [Adapted from Peter Foukal and Judith Lean, Science *247*, 556–558 (1990)]

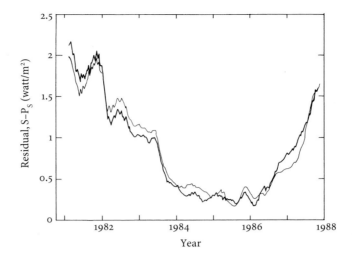

On the short-term scale of minutes to days, sunspot blocking dominates facular brightening; on the long-term scale of months and years, faculae dominate because they have significantly longer lifetimes than sunspots and cover a larger fraction of the solar disc. This may not be the complete story, for the predicted fluctuations in the solar constant often come up short and underestimate the total radiative output at sunspot maximum. We can nevertheless still be certain that the present-day Sun brightens when its magnetic activity increases and it becomes fainter when it settles down.

The observed decade-long, solar-constant variations both warm and cool the planet, potentially producing an equilibrium global temperature change of about 0.2 degrees Celsius. (Because the climate response time is longer than the 11-year solar activity cycle, the actual temperature change caused by the variable Sun is thought to be smaller.) This equilibrium change is comparable to global surface temperature fluctuations during the past half century, suggesting that solar variations, if they are occurring on these longer time scales, could be partly responsible. The Sun's varying radiative output can therefore potentially both lessen and compound global warming, and the Sun must consequently be taken into account when assessing warming caused by human activity (also see Sect. 10.3).

The period of time in which the solar constant has been precisely monitored is not long enough to determine the full scope of its variability. However, indirect historical evidence suggests that the Sun is a likely cause of long-term climate change, as measured in time scales of millennia and centuries, and possibly decades. Variations in the solar constant could effect our environment, climate and humankind. For instance, a prolonged period of missing sunspots between 1645 and 1715 coincided with significant cooling in the northern hemisphere (the Little Ice Age,

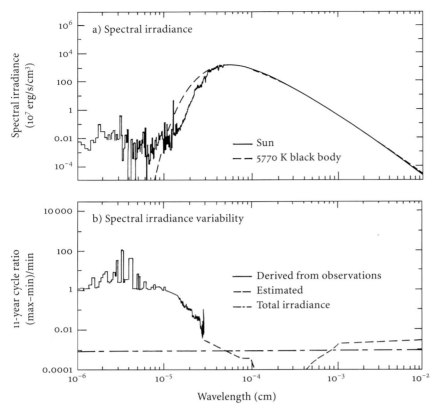

Fig. 9.9. Irradiating Earth. The solar radiative input to the Earth's outer atmosphere at different wavelengths. The Sun's spectrum, or spectral irradiance, at solar minimum is compared with the spectrum of a black body radiator at 5770 degrees Kelvin (*top*). The peak of this curve is located at the visible wavelengths we see with our eyes; the data to the left of the peak are at ultraviolet wavelengths and those to the right correspond to infrared wavelengths. Not shown, at wavelengths longward of the infrared, is the radio portion of the solar spectrum. The bottom curve shows variations in the spectral irradiance from the maximum to minimum of the 11-year activity cycle; it is most pronounced at ultraviolet wavelengths. The solar cycle variations in the spectrally integrated, or total, solar irradiance is indicated by the long, straight dashed line. (Courtesy of Judith Lean, Naval Research Laboratories)

see Sect. 10.6). Observations of stars like the Sun also suggest that larger changes in the solar constant may be possible, being capable of either melting the Earth's ice caps or creating a new ice age, and affecting the future habitability of the planet. Scientists therefore need to watch closely small variations in the solar constant, not really knowing how it will change in the future.

The entire spectrum of the Sun's radiation is modulated by solar activity (Fig. 9.9). Overwhelmingly the greatest part of solar radiation is emitted in the visible and near-infrared parts of the spectrum where the variations are relatively modest. By contrast, there are enormous variations at the short, invisible wavelengths that contribute only a tiny fraction of the Sun's total radiation. The ultraviolet radiation is at least ten times more variable than longer-wavelength visible radiation, so it can contribute significantly to the solar constant variability despite its minor contribution to the total solar radiation itself. The Sun's X-ray emission is even more variable, by at least a factor of one hundred, but it contributes insignificantly to the total.

Solar ultraviolet and X-ray emission is radiated from the hot, outer parts of active regions, in the chromosphere and low corona, and is much more responsive to enhanced magnetic activity than visible emission

Fig. 9.10. Atmospheric Absorption. Most of the Sun's short-wavelength radiation never reaches the ground. The curve shows the altitudes at which about half of a given radiation is absorbed by the Earth's atmosphere. Also shown are the primary atmospheric absorbing species of the radiation within different spectral bands, and the wavelength regions which dominate ozone production and absorption. Fortunately for life, nitrous oxide in the thin atmosphere more than 50 kilometers above the Earth's surface blocks the Sun's highly variable short-wavelength ultraviolet emissions. At lower altitudes, ozone and molecular oxygen absorb longer-wavelength ultraviolet rays, which are also harmful to life. Changes in the Sun's ultraviolet emissions affect the structure of the ozone layer. [Adapted from R. R. Meier, Space Science Reviews *58*, 1–185 (1991)]

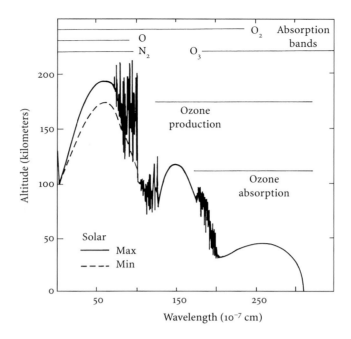

formed in the lower, cooler layers of the solar atmosphere. At times of high solar activity, when flares are more frequent, the solar output of ultraviolet and X-ray radiation can be dramatically enhanced.

Because only the very short wavelengths are energetic enough to ionize the air and transform its chemistry, and because these short-wavelength emissions exhibit the greatest variation, the Earth's upper atmosphere is constantly altered by the Sun's variable, invisible radiation. The shorter-wavelength and more variable solar energy is absorbed at higher altitudes in the terrestrial atmosphere (Fig. 9.10), where the global mean temperature can double between the minimum and maximum of solar activity (Fig. 9.11). The Sun's variable ultraviolet radiation modulates the vital ozone layer at a level comparable to human-induced changes. In contrast, only the least-variable, visible portion of the solar spectrum penetrates through to the relatively placid lower atmosphere, but this sunlight might still vary enough to warm and cool the air.

Thus, fluctuations in the Sun's radiative output can potentially alter global surface temperatures and influence terrestrial climate and weather, alter the planet's ozone layer, and heat and expand the Earth's upper atmosphere. The Sun's cyclic climate warming and cooling, and its stratospheric ozone production and depletion, can mask trends caused by current human activity. To assess human change to date, and fully understand how the atmosphere may respond to future human activity, requires an understanding of solar influences on global change. We must look beyond and outside the Earth, to the inconstant Sun as an agent of terrestrial change.

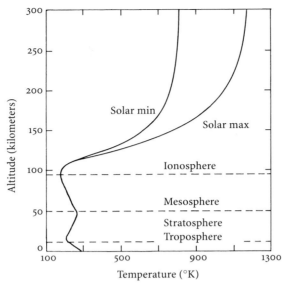

Fig. 9.11. Changes in Solar Heating. The Sun's radiation heats the Earth's atmosphere, creating a layered structure whose boundaries are defined by changes in the temperature profile at approximately 15, 50 and 100 kilometers. At altitudes above 100 kilometers, the increase in temperature with height depends strongly on the 11-year solar activity cycle, becoming greatest at times of maximum activity due to increased ultraviolet and X-ray radiation from the Sun. [Adapted from *Solar Influences on Global Change* (Washington, D. C.: National Academy Press, 1994)]

We therefore next focus our attention on the Earth's protective ozone layer, that is both modulated from above by the Sun's variable ultraviolet output and threatened from below by man-made chemicals.

9.6 SOLAR ULTRAVIOLET AND TERRESTRIAL OZONE

What is ozone and how is it created high in the air? When the high-energy ultraviolet rays from the Sun strike a molecule of ordinary diatomic oxygen that we breathe, they split or dissociate the molecule into its two component oxygen atoms. Some of the freed oxygen atoms bump into, and become attached to, an oxygen molecule, creating ozone molecules that have three oxygen atoms instead of the normal two. These ozone molecules are naturally destroyed when a free oxygen atom combines with an oxygen atom belonging to an ozone molecule, resulting in two molecules of ordinary oxygen, but the Sun keeps on making more ozone. The Sun's ultraviolet rays thereby produce a globe-circling ozone layer high in the stratosphere.

This new molecule absorbs ultraviolet light from the incoming stream of solar radiation. There isn't much ozone high in the air, about one part per million of the gas in the atmosphere, which is roughly equivalent to a drop of wine in a swimming pool, but it can still absorb almost all of the Sun's deadly ultraviolet rays before they reach the ground. (Dangerous ozone is produced by humans close to the ground – see Focus 9A.)

Ozone hasn't always been in the air, for it cannot be formed without oxygen. During the very early days of the Earth's existence, it was too hot for plants to exist and there was probably very little oxygen in the atmosphere. Without any ozone, the Sun's ultraviolet rays penetrated the Earth's primitive atmosphere and played on the surface of the seas, producing the rich supply of organic substances upon which the earliest living organisms most likely fed.

For a few billion years, one-celled algae and bacteria in the sea must have slowly transformed the primitive atmosphere into today's oxygen-rich air. These early organisms consumed carbon dioxide and injected oxygen into the air. In the process, they gradually ruined their own environment and destroyed themselves, since oxygen was a poison to many of them and the Sun's ultraviolet could no longer reach the sea to renew the supply of organic food. Living things nevertheless display a remarkable natural resiliency, for we can still find similar forms of algae and bacteria. Even today, all oxygen breathers have anti-oxidents that keep their cells from being poisoned by oxygen.

The amount of ozone that is now in our atmosphere is subject to the vagaries of the Sun's ultraviolet radiative output. (The active, changing solar atmosphere is responsible for most of the Sun's variable ultraviolet emission.) Ozone molecules are constantly being created by, and destroyed by, chemical reactions that involve ultraviolet sunlight. A balance or equilibrium is struck between creation and destruction of the molecules, but this equilibrium amount varies with the inconstant Sun.

The ozone is enhanced and depleted, and then enhanced again, when the solar atmosphere goes from intense turbulence to relative calm and back to its agitated state during the 11-year cycle of solar activity. When the Sun brightens during active periods, the increased ultraviolet output produces more ozone in the stratosphere, and when our home star is less active and dims, there is less ultraviolet sunlight and smaller amounts of ozone are produced.

Cyclic changes in global ozone abundance, caused by variable solar radiation from above, modulate suspected ozone depletion by chemicals wafting up from below. Analysis of satellite ozone data, in general agreement with current models of ozone production and destruction, indicate that the excess ozone generated during the five-year rise in solar ultraviolet activity can effectively cancel the current global man-made depletion during this period, and that the global five-year ozone decrease caused by reduced solar ultraviolet activity now rivals the global anthropogenic destruction over the same period. The Sun could therefore account for a significant fraction of the global decrease in stratospheric ozone measured by satellites during a period of mostly declining solar activity (1978 to 1985), but solar variability cannot account for the fact that ozone has stayed roughly constant since 1985 while solar activity has recovered to levels equal to or beyond those of 1978.

Only when solar ultraviolet variability is known in detail can we confidently and unambiguously extract the man-made component from the satellite measurements. Scientists can attempt to untangle the solar ozone variations from the depletion caused by human activity by monitoring the Sun's ultraviolet output. Since this radiation can only be observed from above the stratosphere, it may also be useful to obtain radio proxy indicators from the ground (Fig. 9.12).

Ozone is also destroyed by energetic solar particles that vary in step with solar activity. Major solar proton events, associated with great eruptions on the Sun, can initiate chemical reactions that are similar to those caused by ultraviolet sunlight, producing reductions in stratospheric ozone. Solar electrons might also create chemical compounds that can destroy the ozone.

Neither ultraviolet sunlight nor ozone are uniformly distributed over the globe. More ozone is produced at low latitudes in the tropics, for example, where the ultraviolet sunlight is most intense and there is little seasonal variability. At higher latitudes, the ozone levels undergo seasonal variations, but the poleward transport of tropical air partially offsets this imbalance.

Man-made chemicals are also carried toward the polar regions, where they are now literally eating holes in the vital ozone layer. If the chemical build-up continues unabated, it might lead to a disastrous glo-

FOCUS 9A

Bad, Ground-Level Ozone, A Harmful Pollutant

Closer to the Earth, in the air that we breathe, ozone is not so beneficial to life, and is a main ingredient of eye-burning smog that can damage lung tissue and plants. During smog alerts in large cities, such as Los Angeles and mountain-ringed Mexico City, children, elderly people, and jogging enthusiasts are advised to stay indoors because of potential ozone damage to their lungs. Children are especially at risk, because, for their body size, they inhale several times more air than adults, and because they tend to spend more time outdoors.

In contrast to the "good" naturally occurring ozone in the stratosphere, which is created by incoming solar radiation, the "bad", ground-level ozone is produced by humans, primarily from compounds emitted in the exhaust of cars and trucks. Ground-level ozone and smog have also been caused by fires raging across fields and grasslands of Brazil and the savannas of southern Africa, with an intensity comparable to the pollution over industrialized regions in Europe, Asia, and the United States. The "bad" ground-level ozone can also be carried hundreds of kilometers by the wind, so the potential damage is not limited to densely populated and heavily industrialized cities.

Fig. 9.12. Inconstant Spectral Output. Solar radio (*top*) and ultraviolet (*bottom*) radiation vary with the 11-year cycle of solar activity becoming more intense at the time of maximum activity (1980 and 1990) and fainter during the intervening minimum (1986). Radio radiation is not absorbed in the Earth's atmosphere, so the radio data (at 10.7 centimeters wavelength) were obtained using a ground-based telescope (at Ottawa, Canada). Because most of the Sun's ultraviolet radiation is absorbed in our air, these measurements were obtained from the Nimbus 7 satellite in orbit above the atmosphere. [Adapted from R. F. Donnelly, Presented at the Seventh Quadrennial Solar-Terrestrial Physics Symposium in the Hague (1990)]

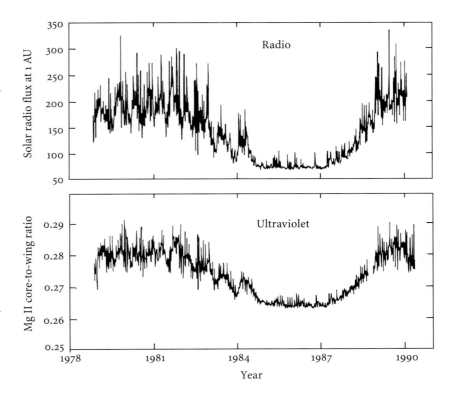

bal depletion of ozone, eventually surpassing any modulation caused by the Sun's variable output. As we shall see, progress has been made in outlawing the ozone-destroying chemicals. Yet, our ability to reliably determine the current ozone depletion, as well as its future recovery, depends on adequate knowledge of how the protective layer is damaged or restored by the Sun. With this caveat in mind, we next discuss the ozone depletion caused by humans and then international efforts to prevent it.

9.7 THE VANISHING OZONE

The amount of ozone in the stratosphere resembles the level of water in a leaky bucket. When water is poured into the bucket, it rises until the amount of water poured in each minute equals the amount leaking out. A steady state has then been reached, and the amount of water in the bucket stops rising and will stay the same as long as you keep pouring water in at the same rate. However, if you pour the water in at a different rate, or punch a few more holes in the bucket, the steady-state level of water in the bucket changes. The amount of ozone in the stratosphere changes in a similar way if there is a variation in the amount of solar ultraviolet streaming into it, or if we punch holes in the ozone layer, as we are now doing with chemicals used in our everyday lives.

A springtime ozone hole has been discovered over Antarctica at the South Pole (Fig. 9.13). It forms within a vast polar vortex that resembles a huge dust devil, the eye of a hurricane, or the whirlpool formed by water draining from a giant's bathtub. Each year the gaping hole opens up during Antarctic spring when the sunlight triggers ozone-destroying chemical reactions, and begins to disintegrate in the early polar fall when the long sunless winter begins. Ozone-depleted air is dispersed globally, and the ozone is slowly restored to fill in the hole until the cycle repeats in the following year.

The ozone layer is itself invisible, so you can't see the ozone hole. Instruments called Dobson spectrophotometers, placed on the ground, measure the amount of ultraviolet radiation getting through the atmosphere at specific wavelengths or spectral lines. The absorption lines are strengthened when there is more ozone, and weakened if there are lesser amounts. Other instruments aboard orbiting satellites now use backscattered ultraviolet spectral techniques to remotely sense the amount of ozone from above, providing a global picture of ozone concentrations.

A small springtime decrease of ozone above the South Pole was first detected in 1957–58, during the International Geophysical Year, using the instrument that now bears Dobson's name. (As discussed later, the ozone depletion above Antarctica has subsequently grown in size.) British scientists had previously found that the amount of ozone varies annually throughout Europe. When this annual variation was compared to that measured at the Antarctica station, assuming a six-month difference in the seasons, the ozone values for polar spring were much lower than was expected. It was initially thought that some large mistake had been made, or that the instrument had developed some major fault, but in the polar fall the ozone values jumped to those expected from extrapolation of the European results. According to G. M. B. Dobson's historical account, written in 1968:

> It was not until a year later, when the same type of annual variation was repeated, that we realized that the early results were indeed correct and that Halley Bay showed a most interesting difference from other parts of the world. It was clear that the winter vortex over the South Pole was maintained late into the spring and that this kept the ozone values low. When it suddenly broke up in November both the ozone values and the stratosphere temperatures suddenly rose.[45]

The total ozone concentration was measured at Halley Bay, Antarctica for several decades, always showing an anomalous springtime loss that became steadily larger as the years went on. The Antarctic ozone hole eventually became so dominant in the ground-based data that it could not be ignored, and astounded space-age scientists rechecked their satellite data, finding that they had dutifully and unwittingly recorded the unexpected ozone collapse for several years. The embarrassed experts had

Fig. 9.13. Hole in the Sky. A satellite map showing an exceptionally low concentration of ozone, called the ozone hole, that forms above the South Pole in the local spring (in October 1990); it has an area larger than the Antarctic continent (shown in outline below the hole), a land area covering approximately 14 million square kilometers. Eventually spring warming breaks up the polar vortex and disperses the ozone-poor air over the rest of the planet, but some of it remains over the South Pole until the following winter to accelerate the cooling. (Courtesy of NASA, Science Source/Photo Researchers)

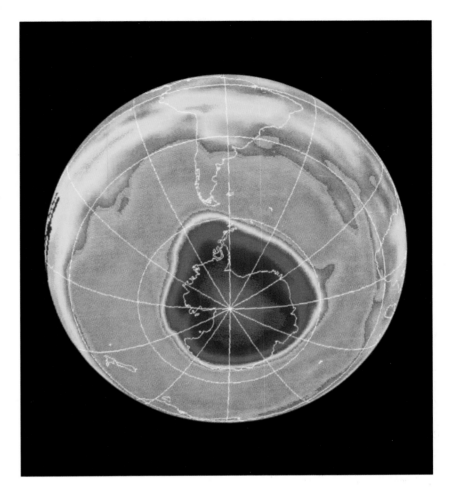

programmed their computers to reject measurements that diverged sharply from normal conditions, so the now-famous ozone hole had been discarded as an anomaly.

The Antarctic ozone hole has grown in extent year by year for more than a decade, and shows no signs of abating. The hole is now about as wide as the North American continent and nearly as deep as Mount Everest is tall. (In a six-week period during the southern spring, effectively all the ozone is destroyed at altitudes between about 10 and 20 kilometers.)

So what's destroying the ozone layer that protects life on Earth from excessive ultraviolet radiation? Mario Molina and F. Sherwood Rowland, two chemists from the University of California at Irvine, fingered the culprit with prophetic insight in 1974. As they expressed it:

Chlorofluoromethanes are being added to the environment in steadily increasing amounts. These compounds are chemically inert and may remain in the atmosphere for 40 to 150 years, and concentrations can be expected to reach 10 to 30 times present levels. Photodissociation of the chlorofluoromethanes in the stratosphere produces significant amounts of chlorine atoms, and leads to the destruction of atmospheric ozone.[46]

So, the ozone layer is being wasted away by synthetic chlorine compounds, released into the atmosphere by human activity. They are man-made gases, invented about a half-century ago and now given the euphonious name "chlorofluorocarbons". This name is a giveaway to their composition, chlorine, fluorine, and carbon, and is often abbreviated as CFC. (A number sometimes follows the shorthand CFC notation, indicating the number of chlorine, fluorine, carbon and hydrogen atoms in each molecule, the most widely used being CFC-11 and CFC-12.)

The CFCs are entirely of human industrial origin with no counterparts in nature, and for several decades they were hailed as wonder chemicals. Their molecules are assembled using the strongest chemical bonds permitted by nature, making them almost impervious to breakage. They are very stable, nontoxic, noncorrosive, nonflammable, and relatively easy to manufacture. The CFCs have been widely used as coolants in refrigerators, freezers and air conditioners, as foaming agents for insulation found in coffee cups and packaging, as propellants in hair spray and deodorants (now banned in the United States), and as cleaning solvents in the process of manufacturing computer chips.

Although tens of millions of tons of CFCs have been released into the air, their combined concentration isn't very significant, only about one CFC molecule for every 2 billion molecules in the air, but even these seemingly insignificant amounts can have enormous impact.

When initially released in the lower atmosphere, the hardy chemicals don't interact chemically to form other substances that would get removed from the atmosphere naturally. And they're not soluble in water, so they don't get rained out. They are so inert and stable that once released into the atmosphere the CFC molecules can survive for more than a century, permitting them to drift and waft up into the ozone layer.

In the lower atmosphere, CFCs are protected from the Sun's intense ultraviolet radiation by the ozone layer, enabling them to migrate intact into the stratosphere. Once there, however, the energetic solar ultraviolet tears the sturdy CFC molecules apart, producing free chlorine atoms which in turn destroy the ozone. As Molina and Rowland theorized more than two decades ago, the chlorine is the real culprit in the destruction of atmospheric ozone, being able to eat up thousands of times its own weight in ozone.

Once freed in the stratosphere, a chlorine atom can react with an ozone molecule, taking one oxygen atom to form chlorine monoxide; the ozone is thereby returned to a normal oxygen molecule and its ultraviolet absorbing capability is largely removed. If this was all that happened, there wouldn't be much concern. However, when the chlorine monoxide encounters a free oxygen atom, the chlorine is set free to strike again and again. Because the cycle repeats over and over, a little chlorine goes a long way. In technical terms, the chlorine acts as a catalyst, initiating a series of chain reactions that destroy ozone, while surviving the process. (The catalytic converter on your car acts in a similar way, catching anything that passes through the engine unburned.)

A single chlorine atom can disrupt as many as 100 000 ozone molecules before it is captured and locked away in some other molecule. A small amount of chlorine can therefore produce significant changes in the ozone layer, which is itself very rarefied and accounts for only about one-millionth of the molecules in our own air.

In places, chlorine can destroy ozone at a far faster rate than the gas is replenished naturally. The ozone loss then exceeds the ozone creation, as if water were being poured into a bucket with its bottom cut out. The Sun's powerful ultraviolet rays can then penetrate to ground, producing widespread damage to living things.

Why doesn't chlorine keep on consuming ozone until there is none left in the stratosphere? The cycle of destruction is eventually broken when a chlorine atom, instead of interacting with ozone, interacts with methane in the atmosphere to form hydrochloric acid. The acid diffuses downward from the stratosphere to the troposphere, and because it's soluble in rain the chlorine is finally washed out, closing the cycle.

The simple scenario of ozone-destroying chlorine, released from CFCs by ultraviolet sunlight, does not explain why severe ozone depletion is limited to the Antarctic region or why it is observed only in the spring. Something must be focusing and intensifying chlorine's destructive power both in space and time.

Although the CFCs are mainly released by human activities in the Northern Hemisphere, where the world's industries are concentrated, the chemicals are redistributed by global winds. Indeed, substantial amounts of ozone are found over each pole of the Earth during the long polar night, despite the complete absence of sunlight, because stratospheric winds transport ozone into the polar regions. Each sunless winter, steady winds blow in a circular pattern over the ocean that surrounds Antarctica, trapping a huge air mass inside for months at a time. The whirling winds concentrate the ozone-destroying chemicals within a vast, towering polar vortex, confining the most serious ozone depletion to regions above the South Pole and isolating it from the rest of the world for months at a time (also see Focus 9B).

An Arctic vortex, about the size of Asia, forms over the North Pole in the northern winter, although weaker, more unstable and warmer than the southern one. While the Antarctic is land surrounded by ocean, the Arctic is ocean surrounded by land, which directs warm air northward before the Sun's rays arrive. The huge ice-covered land mass in Antarctica is therefore much colder than the Arctic ocean or land, and this accounts for the formation of the much stronger Antarctic vortex and its polar stratospheric clouds.

Still, an ozone hole could be coming closer to home; one might even be forming right over your head. Already, even without a northern ozone hole the stratospheric ozone has been depleted by almost 10 percent in just four decades, at least in the winter and early spring.

Instruments aboard NASA's Upper Atmospheric Research Satellite (UARS) show high concentrations of ozone-eating chlorine monoxide within the Arctic vortex during the northern winter when it is cold enough for some polar stratospheric clouds to form. The Arctic vortex is then chemically primed for ozone depletion with levels of reactive chlorine as high as those observed six months previously during the south-

FOCUS 9B

High-Altitude Ice Clouds

The ozone hole also requires the presence of polar stratospheric clouds of ice crystals. Whipping around the pole, the high-speed vortex winds shut out air from the warmer equatorial regions and keep the temperatures very cold inside. During the long, dark winter, the stratospheric air over Antarctica is therefore colder than elsewhere, allowing ice crystals to form even though the amount of water is very low.

Fortunately, over most of the year and throughout much of the stratosphere, ozone-destroying chlorine atoms released from CFCs remain sequestered and locked up in relatively inert reservoir molecules (such as hydrogen chloride and chlorine nitrate) and cannot immediately attack ozone. However, the ice clouds convert the benign and inert reservoir molecules to more reactive forms, ultimately forming chlorine monoxide which can efficiently destroy ozone when sunlight is present.

The ozone-destroying reactions begin in earnest when the Sun rises over the South Pole, after the long winter night. When the first rays of that long-awaited dawn strike the ice crystals, they accelerate and facilitate the wholesale destruction of the ozone until virtually all of it has been eaten up. Then, with the onset of warmer weather, the polar vortex loses its integrity and breaks up, diluting the rest of the world's atmosphere with ozone-poor air. (The ice crystals provide surfaces or platforms on which chlorine and ozone can alight and interact more readily than if they were free.) Such processes have been confirmed by an observed rise in the ozone-killing chlorine monoxide over the South Pole during the formation of the ozone hole and the subsequent disappearance of chlorine monoxide when the ozone level returns to normal.

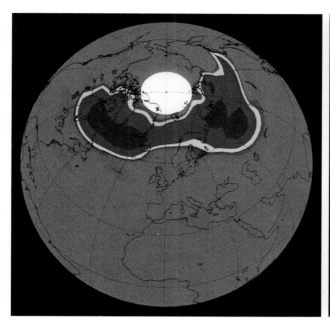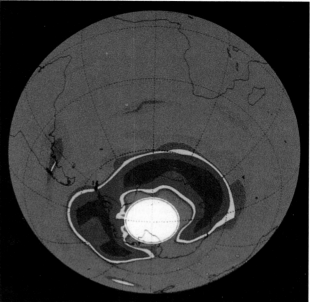

Fig. 9.14. Ozone-Destroying Chlorine in the Stratosphere. Red and darker colors indicate chlorine monoxide (ClO) abundances that are great enough to destroy ozone molecules at the rate of 1 percent per day or greater; the white circles at the poles are where no measurements were made. The map on the right shows ClO above Antarctica in August, 1992, while the left map shows its distribution in the northern hemisphere in February, 1993. Dangerous abundances extend over populated areas of northern Europe, Scandinavia and Russia. These maps were produced by the Microwave Limb Sounder aboard the Upper Atmosphere Research Satellite. (Courtesy of Joe W. Waters, Jet Propulsion Laboratory)

ern winter in the Antarctica vortex (Fig. 9.14). If the temperatures had remained cold enough, northern spring sunlight and the chlorine monoxide could have combined to create an Arctic ozone hole over populated areas of Europe and Canada.

Chemicals have led to ozone depletion of at least 3 percent worldwide over the past decade, including the mid-latitudes where most people live. The varying solar ultraviolet modulates the global ozone layer at this level, but eventually chemical ozone depletion by human activity will dominate any solar effect. So there is plenty of cause for alarm if these chemicals continue to be produced.

9.8 HEALING THE PLANET

What's all the fuss about anyway? Who cares if chemicals are punching a few holes in the sky and letting a little more sunlight reach the ground?

The ozone layer normally prevents an especially dangerous form of solar ultraviolet radiation, called UV-B, from reaching the ground. It has relatively short wavelengths between 280 and 320 nanometers (a nanometer is one ten-millionth of a centimeter). Longer ultraviolet wavelengths reaching the Earth's surface (UV-A at 320 to 400 nanometers) are less effective in producing biological changes. When a DNA double helix in the nucleus of a cell is exposed to UV-B radiation from the Sun, a single strand can bond to itself, becoming balled up enough to prevent replication. This will reduce the effectiveness of the immune system of animals and humans, increasing their vulnerability to infections and diseases such as cancer.

The virulent UV rays can also make the lens of the eye cloud up with cataracts, which bring on blindness if untreated. Recently, UV-B light from the Sun has been identified as the specific cause of the genetic defect that produces a common skin cancer. (The UV-B cripples a gene named p53, which normally prevents riotous cell growth, leading to squamous-cell carcinoma.)

Perhaps the danger closest to the human heart, and the one most responsible for rapid political action, is the threat of a world-wide epidemic of skin cancer. Overzealous sunbathing has, in fact, long been associated with increased skin cancer for fair-skinned people. The U. S. Environmental Protection Agency has, for example, predicted that in the United States alone 12 million people will develop skin cancers by the year 2050 because of reduced ozone, and that 200 000 of them will die.

The thinning of the ozone layer may upset, perhaps irrevocably, the fragile balance of life at sea. Increased UV-B can damage or even wipe out one-celled plants, called phytoplankton, which live near the surface of the sea and are the base of the ocean food chain. (Satellite images show carpets of phytoplankton spanning the surface of the world's oceans in local spring – see Fig. 9.3; the name comes from the Greek words *phyto* for plant and *plankton* for drifting.) When the plankton population declines, so will the populations of fish that eat plankton, and the mammals that eat the fish – seals, dolphins, whales, and humans.

The phytoplankton are not defenseless. They have colored pigments that they deploy like sunglasses to absorb the UV-B and protect their internal mechanisms, but their photosynthesis still appears in some studies to have decreased as the result of enhanced UV-B. Recent investigations nevertheless suggest that the algae might thrive in the presence of increased ultraviolet, because the lethal rays are even more harmful to the organisms that feed on them.

Humans can cover themselves, stay indoors, or acquire darker skins. After all, people near the equator usually have a darker color, which tends to protect them from skin cancer in the intense tropical sunlight. However, at some point living things, from phytoplankton to human beings, may no longer be able to cope with the increasing ultraviolet. The protection of the ozone layer could simply diminish too rapidly for many species to adapt quickly enough to survive.

These concerns, together with the now-famous Antarctic ozone hole, sparked public awareness of the fragile ozone layer and served as a powerful political stimulus to limit and eventually ban the production of CFCs. Faced with the evidence of the vanishing ozone, international diplomats signed a treaty, known as the Montreal Protocol, in 1989, agreeing to a 50 percent reduction in the CFCs by the turn of the century. The protocol was strengthened a year later when the diplomats agreed to a complete phase-out of the ozone-destroying chemicals in the industrialized countries by the year 2000 and in the developing countries by 2010. When it was subsequently shown that the erosion of the ozone layer is accelerating at a faster rate than predicted, worried leaders of 128 countries agreed to an extended Montreal Protocol to eliminate the most important ozone-destroyers, such as CFC-11 and CFC-12, by the end of 1995. The number of participating countries, as well as the list of banned chemicals, is growing all the time. (This agreement was undoubtedly eased by the development of substitutes for CFCs in refrigeration, air conditioning, foaming and cleaning.)

However, even if humans stopped putting CFCs into the air today, the damage could not be undone. Because of their long lifetime and slow diffusion into the stratosphere, the presence of CFCs in the ozone layer would continue unabated for almost half a century if we stopped producing them now. Many of the ozone-destroying culprits have already been dumped into the atmosphere, at least 20 million tons of them. The worst of the ozone destruction caused by human activity will probably come around the end of the twentieth century, when the peak load of the ozone-ravagers, now slowly wafting upwards, reaches the stratosphere. After that, ozone loss should be stabilized and the trend reversed as the chemicals are gradually removed by slow washing-out processes.

The ozone layer will not regain full strength until well into the latter half of the twenty-first century when it should then recover to the natural levels that existed before the ozone hole was discovered. In the meantime, scientists will continue to monitor the ozone layer using a series of NASA satellites; but unfortunately there is no good plan to measure the varying solar ultraviolet or extreme ultraviolet radiation. Measurements of the varying solar ultraviolet output are required to determine whether, and to what extent, human activities are changing, or will change, the world's climate and temperature. We therefore now turn our eyes up to the sky, whose color and energy are supplied from above by the Sun and whose temperature may be altered by human forces from below.

The Slave Ship, 1840. In this painting, by Joseph M. W. Turner, the setting Sun illuminates a trough between two enormous swells in the storm-tossed sea. The dark, indefinite human forms are found in foam and whirling water. They are sick and dying slaves thrown overboard when an epidemic broke out, so the slavetrader could say they were lost at sea and claim insurance. (Courtesy of the Museum of Fine Arts, Boston, Henry Lillie Pierce Fund)

CHAPTER 10

Fire and Ice

10.1 CLEAR SKIES AND STORMY WEATHER

Look up! The clear air turns into the blue sky! This is because air molecules scatter blue sunlight more strongly than other colors.

The night sky is black because there is no sunlight to illuminate the air. And the sky on the Moon is black even when the Sun shines on it, for the Moon has no atmosphere to scatter sunlight.

When the Sun is high in the sky it is colored yellow. At sunset the Sun's rays pass through a maximum amount of atmosphere; most of the blue light is then scattered out of our viewing direction, and the setting Sun is colored red. Dust in the air also helps redden sunsets.

Over the eons, our planet has been shaped and re-shaped by nature's powerful forces, and "air-conditioned" by several types of independent cycles. The most obvious of these cycles is the water cycle that brings us our daily weather and produces the climatic differences that starkly distinguish one part of the globe from another.

The water cycle begins in sunlight. About one-third of the solar energy reaching the Earth's surface is expended on the evaporation of sea water. This evaporation releases warm fresh-water moisture into the air and cools the surface of the ocean. The moisture rises high into the cold atmosphere, where clouds are formed. Winds drive the clouds for great distances over sea and land. Rain or snow from the clouds can then fall to land, refreshing streams, lakes and underground reservoirs. The water then runs down to the sea, where the cycle begins once more.

All of the water in the oceans passes through this water cycle once in 2 million years. Yet, the ocean waters are at least 3.5 billion years old, so they have, on average, completed more than a thousand such cycles.

Our weather is largely a consequence of this water cycle, driven by winds and currents that are themselves generated by the unequal distribution of sunlight on the globe. Tropical regions of the Earth receive the greatest amount of heat because the Sun's almost vertical rays travel to the ground through the least amount of intervening air. At higher latitudes nearer the poles, sunlight strikes the ground at a glancing angle and it must also penetrate a greater thickness of absorbing atmosphere. As the Sun heats the land, sea and air, warm winds and ocean currents attempt to correct the imbalance, carrying heat from the equator to the poles, or from the warmer to colder places, thus reducing the initial temperature difference, while cold winds and water move in the opposite direction.

John Updike has poetically captured this stately Sun-driven circulation of air and water:

> All around us, water is rising
>   on invisible wings
> to fall as dew, as rain, as sleet, as snow,
>   while overhead the nested giant domes
> of atmospheric layers roll
>   and in their revolutions lift
>   humidity north and south
> from the equator toward the frigid, arid poles,
>   where latitudes become mere circles
> Molecular to global, the kinetic order rules
>   unseen and omnipresent ... .[47]

Exactly how the global weather patterns develop depends on the long-term changing pattern of land and sea. In another of nature's cycles, the great land masses on Earth are continually reorganizing, growing in places and eroding away or breaking apart in others. Over long periods of time, measured in hundreds of thousands of years, entire continents and oceans can be destroyed or created anew, remodeling the entire surface of the Earth. This restlessness influences Earth's climate and weather, guiding a changing flow of air and ocean currents that create entirely different global weather patterns.

Nowadays, and in all former times, it is the Sun-driven seasons that dominate our weather. As we all know, the Earth is irrevocably bound in orbit about the Sun in a cosmic dance by the invisible force of gravity. During a year, the Sun appears to be carried up and down in the sky because the Earth's rotational axis is tilted by 23.5 degrees with respect to the plane of its orbit about the Sun. In each hemisphere, the greatest sunward tilt defines the summer part of its orbit when the Sun is more nearly overhead and its rays strike the surface more directly than in the winter part when that hemisphere is tilted away from the Sun.

This annual change in solar radiation produces large seasonal temperature fluctuations in the northern hemisphere where most of the world's land is now found. The difference in surface air temperature, averaged over the northern hemisphere, between winter and summer, is 15 degrees Celsius. This seasonal temperature change is several times larger than the average surface temperature change between the last ice age and now. In the southern hemisphere, the Sun-driven seasonal temperature response is about one third of that in the north, because of the larger amounts of ocean in the south.

The varying amounts of solar radiation that reach the Earth's surface should continue to dominate its temperature and the weather for years to come, with modifications caused by our restless Earth. Without both sunlight and our thin atmosphere, there would be no blue skies or red sunsets, no clouds or rain, and no climates or seasons. But now a new force is changing the air, and that force is us. It may eventually transform our climate.

10.2 TURNING ON THE HEAT

Over the past century, the planet as a whole has warmed up in fits and starts by almost half a degree Celsius. However, the global temperature curve fluctuates between warm and cool periods, so the trend is very sensitive to the beginning and end points of the record (see Fig. 10.1). The world heated up by about half a degree Celsius between 1920 and 1940, for example, but it entered a cooling phase after that. Fluctuations in the global temperature record prohibit the detection of a smooth, monotonic increase expected from greenhouse warming, but the overall global temperature, averaged over long time intervals, certainly has not decreased during the twenty-first century. (These temperature trends are also averaged over the entire globe, with some parts of the world warming more then the average and others cooling.)

From 1940 to 1970 temperatures dropped so much that some experts predicted a coming ice age. In subsequent decades temperatures began to rise again (also see Fig. 10.1), and the world became unusually hot on average. Recent long, hot summers might mark the beginning of a pronounced warming trend or they could be part of a random and reversible change, a natural fluctuation in the global temperature curve.

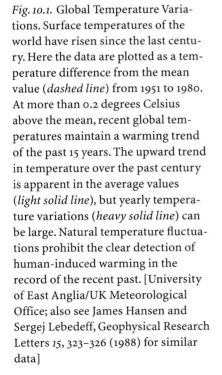

Fig. 10.1. Global Temperature Variations. Surface temperatures of the world have risen since the last century. Here the data are plotted as a temperature difference from the mean value (*dashed line*) from 1951 to 1980. At more than 0.2 degrees Celsius above the mean, recent global temperatures maintain a warming trend of the past 15 years. The upward trend in temperature over the past century is apparent in the average values (*light solid line*), but yearly temperature variations (*heavy solid line*) can be large. Natural temperature fluctuations prohibit the clear detection of human-induced warming in the record of the recent past. [University of East Anglia/UK Meteorological Office; also see James Hansen and Sergej Lebedeff, Geophysical Research Letters *15*, 323–326 (1988) for similar data]

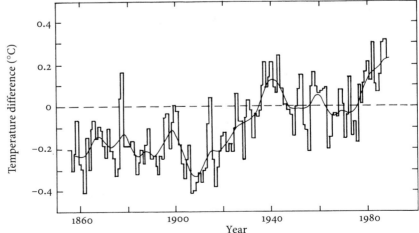

There can be no doubt that we are polluting our air, experimenting with the whole Earth without knowing the outcome. Since the industrial revolution, humans have released carbon dioxide and other heat-trapping, greenhouse gases into the atmosphere at an ever increasing rate. Around the globe, cars and factories belch huge amounts of the potentially dangerous gases into the air, and no end to the buildup is in sight.

Already in 1957 Roger Revelle and Hans Suess noticed, with prophetic insight, that the climate might be drastically changed if we keep on altering the composition of our air, stating that:

> During the next few decades the rate of combustion of fossil fuels will continue to increase, if the fuel and power requirements of our world-wide industrial civilization continue to rise exponentially ... . Thus human beings are now carrying out a large scale geophysical experiment of a kind that could not have happened in the past nor be reproduced in the future. Within a few centuries we are returning to the atmosphere and oceans the concentrated organic carbon stored in sedimentary rocks over hundreds of millions of years.[48]

The increase in carbon dioxide has been measured without stop from 1958 to the present day in the clean air at the Mauna Loa Observatory on top of a volcano in Hawaii. (The measurement site is far from cars, people, towns, and other localized sources of pollution.) The most striking aspect of the carbon dioxide curve (Fig. 10.2, top) is its smooth, systematic increase over the entire period of observations. Since 1958, atmospheric concentrations of carbon dioxide have increased from 315 parts per million per volume to 360 parts per million in 1994. That's an increase of 13 percent in just 35 years, and both the increase and its acceleration have been rock steady – as inexorable as the expansion of the world's population, human industry and pollution.

Moreover, we now know, from ice core studies, that the buildup of carbon dioxide has been going on since the beginning of the industrial revolution in the mid-eighteenth century (Fig. 10.2, bottom). Air is sealed off in bubbles when the ice is laid down, and can be extracted from cores drilled deep within layered glaciers. In the past century, a mere blink in the eye of cosmic time, there has been a dramatic rise in the amount of heat-trapping carbon dioxide in the air, largely because of the burning of fossil fuels, such as coal, oil, and gas. (The torching of forests to increase tillable land, at the present global rate of 1.5 acres each second, now accounts for about one quarter of the rise.)

There are now about 750 billion tons of carbon in the air in the form of carbon dioxide, and this amount is now increasing at the rate of approximately 3 billion tons each year. (The amount released by human activity each year is 7 billion tons; only about half of it remains in the atmosphere.) In other words, each person now on Earth is, on average, dumping about a ton of carbon dioxide into the air every year.

Fig. 10.2. Rise in Atmospheric Carbon Dioxide. The plot shown at the top gives the average monthly concentration of atmospheric carbon dioxide in parts per million of dry air versus time in years observed continuously since 1958 at the Mauna Loa Observatory, Hawaii. It shows that the principal waste gas of industrial societies, carbon dioxide, has risen steadily during the past forty years. The fluctuations are due to seasonal variations; summertime lows are caused by the uptake of carbon dioxide by plants. The other figure (*bottom*) shows how the Mauna Loa atmospheric data (*solid circles*) relate to the carbon dioxide ice core data (*open circles*), indicating an exponential increase in the amount of carbon dioxide over the past two and a half centuries. This diagram also provides a connection with temperature and carbon dioxide changes recorded in ice cores during past ice ages – see Fig. 10.5. (Courtesy of Charles D. Keeling, Scripps Institution of Oceanography, University of California, San Diego)

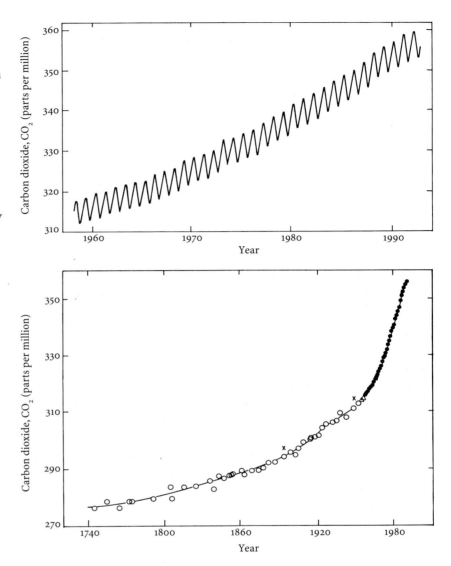

Climatologists are, to put it mildly, concerned about the possibility of global warming by the greenhouse effect if this irreversible buildup of carbon dioxide continues unabated. There is general agreement that the amount of carbon dioxide in the air will double by the middle of the twenty-first century without remedial action, and that the heat-trapping gas will eventually raise the temperature of the Earth's surface if it continues to accumulate. About a century ago, for example, the Swedish scientist Svante Arrhenius estimated how much heating would result when the atmospheric carbon dioxide was doubled, obtaining a surface temperature rise of about 5 degrees Celsius (also see Sect. 9.2).

Scientists now use sophisticated computer models, known as general circulation models, to forecast the amount of climate warming that will result from increased greenhouse gases such as carbon dioxide. These models require vast amounts of computer time on powerful and expensive super-computers, and this work is now carried out at national centers throughout the world. They predict a rise in the global surface temperature of between 1.5 and 4.5 degrees Celsius when the greenhouse gases double; the largest predicted rise does not differ significantly from Arrhenius' estimates a century ago.

During the last few decades, other heat-trapping gases, as well as carbon dioxide, have been accumulating noticeably in the atmosphere (see Focus 10A). They now contribute about as much to global warming as carbon dioxide. However, industrial and volcanic emissions of sulfur form small particles, called sulfate aerosols, that may be reflecting solar radiation back into space, thereby masking the greenhouse effect over some parts of the Earth (see Focus 10B).

Critics of the global warming hypothesis say that the computer models cannot be trusted because of the massive uncertainties (see Focus 10B), and because they fail to predict the exact pattern of warming and cooling actually experienced over the last century. Forecasts based upon extrapolations and unverified models are indeed highly suspect, particularly when they often don't take into account varying external factors such as the Sun's changing radiative output. It may well be that the world's climate, with its many variables and nonlinear feedback mechanisms, is just too chaotic, complex and unpredictable for an accurate description.

Large natural fluctuations, superposed on the record of global temperature changes, mask our ability to clearly detect the relatively small man-made greenhouse effect. During the past century, the Earth's surface temperatures have not increased in a steady or inexorable way, as would be expected by the continued increase in carbon dioxide over this period (see Fig. 10.1). The drop in temperature between 1940 and 1970 is particularly difficult to explain as a greenhouse phenomenon, because this entire interval was characterized by strong economic growth, increasing emission of greenhouse gases, and supposedly increasing temperatures. However, recent model investigations indicate that sulfate aerosols and ozone depletion may account for at least part of this discrepancy (see Focus 10A and Focus 10B).

Forecasts of future global warming are likely to remain hampered by such limitations, at least for the coming decade or two. Perhaps that is not surprising when you consider that less complex computer programs used to forecast the local weather are not all that reliable. Even with daily satellite images of weather pattern, predicted winter storms in the northeastern United States fail to occur, and hurricanes often veer outside their predicted path.

Still, some climate experts suggest that we are now trapped beneath a suffocating tent of gases that is slowly cooking the planet. For instance,

FOCUS 10A

New Heat-Trapping Gases in Our Atmosphere

Increased carbon dioxide may be responsible for only about half the "unnatural" greenhouse effect. The other half of the warming may be due to chlorofluorocarbons (CFCs), methane, and nitrous oxide.

The addition of one CFC molecule can have the same greenhouse effect as the addition of 10 000 molecules of carbon dioxide to the present atmosphere, but fortunately the warming effect of these industrial chemicals may soon be leveling off since they have been banned on the basis of their ozone-destroying capability. (Contrary to popular misconceptions, the thinning of the ozone layer does not by itself make the Earth's surface hotter.) Moreover, certain models suggest that the ozone depletion caused by CFCs may cool the surface temperature, perhaps countering some of the CFC contribution to greenhouse warming. (Ozone is a greenhouse gas, so you get less warming when chemicals destroy it.)

Like carbon dioxide, the atmospheric concentration of methane, also known as natural gas, has more than doubled since the start of the industrial revolution. Although atmospheric methane remains about 100 times less abundant than carbon dioxide, each incremental molecule of methane is about 20 times more effective in trapping heat than each additional molecule of carbon dioxide. Methane is produced by some of the bacteria that thrive in oxygen-free places like swamps, rice paddies, landfills, garbage dumps, termite colonies, and in the stomachs of cows. (The United States Environmental Protection Agency has recently spent half a million dollars to find out if cattle belch enough methane to contribute to global warming.) Methane in swamps, known as marsh gas, sometimes ignites spontaneously, producing flickering blue flames called will-o'-the-wisps. Some methane also escapes from coal mines, natural gas wells and leaky pipelines.

Nitrous oxide, or laughing gas, is also building up in the air, although not as rapidly as methane. The current rate of increase is about 0.2 percent a year, primarily as the result of nitrogen-based fertilizers but also from burning of fossil fuels in cars and power plants.

FOCUS 10B

Limitations to Current Models of Global Warming

Computer models of global warming now represent a stripped-down version of the Earth's real climate. Approximation techniques required to solve the equations with available computing resources involve finite grids that are much larger than the scale of variation of critically dependent variables. The number of points is so low that they blur distinctions between land and sea, mountains and plains, or clouds and ice; one point on the grids represents average conditions over entire countries and the models cannot forecast localized conditions within them.

A major uncertainty in forecasting global warming is the indefinite knowledge of the future fate of heat trapping gases. Right now, about half the carbon dioxide released in the atmosphere by human activity does not stay up in the air. Something is removing the waste gas, thereby cushioning us from the full brunt of global warming. The Earth has two carbon-dioxide lungs, the forests and oceans, but they apparently do not account for all the missing carbon. Some researchers think that the regrowth of northern-hemisphere temperate forests may be soaking up the excess.

There are even greater uncertainties involving nonlinear, interactive feedback mechanisms that can either amplify (positive feedback) or reduce (negative feedback) small changes in climatic variables. An example is the ocean temperature. Water absorbs more carbon dioxide when it is colder, which is why you should always keep a carbonated beverage cold. When the oceans get warmer, they will absorb less carbon dioxide, leaving more of it in the air, strengthening the greenhouse effect, and further warming the oceans in a positive feedback.

Sulfur, produced when burning coal and oil, may lead to a negative feedback as the result of sulfate aerosols that scatter incident sunlight and cool the Earth. Volcanoes also inject significant amounts of sulfate aerosols in the air, and that shows up in the global temperature record.

Water vapor plays a critical role in warming due to the greenhouse effect. Like carbon dioxide, water vapor is effective in trapping heat, and without moisture in the air there would be less global warming. As increasing temperatures evaporate more water from the oceans, the additional water vapor in the atmosphere and possible change in cloudiness will alter the greenhouse effect.

Because we don't understand clouds, the future of global warming is also cloudy. When the Earth warms it should produce more clouds along with the increasing water vapor. However, when the clouds are included in the computer models, the agreement among them varies widely. Some models forecast global cooling by clouds; they reflect more incident sunlight into space, lessening the amount that warms the ground in a negative feedback. Other models predict that clouds block the flow of heat from the Earth's surface, thereby enhancing global warming in a positive feedback. It's just a question of which effect dominates and how strong it is, but that's the difficulty, for no one seems to know for sure. Clouds cover roughly half the area of the Earth at any given time, so this is not a negligible uncertainty.

James E. Hansen, director of NASA's Goddard Institute for Space Studies in New York City, asserted in 1989 that:

> Since the global climate models indicate that the increase in greenhouse gases [in the past century] should have increased the temperature by 0.5 degrees Celsius, the modeling and observational evidence are consistent in indicating that the ["unnatural"] greenhouse effect is changing our climate.[49]

Some would say that he was overstating the case.

We must place the Earth within a broader context and understand the ever-changing Sun if we are to fully assess global warming. The Sun is the driving force for all climate and weather on Earth; yet, it is likely that many climate models do not adequately include effects associated with solar radiation changes. Indeed, as we shall see, global temperature changes during the past century are correlated with variations in solar activity, and the observed temperature variations may have less to do with human activity than was previously thought.

10.3 SOLAR RADIATION AND GLOBAL WARMING

The Sun energizes our climate and weather. It provides the heat that evaporates the ocean water and that produces the temperature variations that drive winds in the air and currents in the sea. And since the Sun is a variable star, you might expect that its changing radiative output would produce global climate change. Yet, until recently the total energy radiated by the Sun was generally considered, for the purpose of climate forecasts, to be constant.

We now know that the ultraviolet and X-ray emissions from the Sun vary greatly, and that the Earth's upper atmosphere acts as a sponge, soaking up the unseen radiation. At times of high solar activity, the Sun pumps out much more of the invisible rays, the air absorbs more of them, and our upper atmosphere heats up; when solar activity diminishes the high-altitude air absorbs less and cools down (also see Sect. 9.5). The resultant high-altitude temperature variations produce cyclic, decade-long changes in stratospheric winds that could produce weather changes in the lower atmosphere by some as-yet-unknown mechanism (see Focus 10C).

Changes in the Sun's visible radiation pass right through to the ground, providing a direct warming or cooling of the lower atmosphere. The visible fluctuations can either lessen or compound global warming caused by atmospheric pollution. Present-day climate models indicate that the observed 0.1 percent variation in the total solar radiative output could produce a global equilibrium temperature change of 0.2 degrees Celsius if it persisted for 50 to 100 years. The same models indicate that

The Sun and Terrestrial Weather

Until recently, correlations between the varying Sun and our weather were treated with about as much skepticism as invasions by extraterrestrials. Now, after centuries of controversy, scientists are finding apparently indisputable links between the Sun's 11-year activity cycle and changes in terrestrial weather. In 1987, for example, the German meteorologist Karin Labitzke and her American colleague Harry Van Loon showed that high-altitude winds and some weather changes may beat to a solar rhythm. They presented convincing evidence that northern winter storms match both the period and phase of the solar cycle.

That is, the stratospheric wind that swirls above the North Pole, and mid-winter warming in the United States and Western Europe, are correlated remarkably well with the Sun's 11-year activity cycle for the past four decades, provided that a switch in the direction of equatorial stratospheric wind every two years is taken into account. (The equatorial stratospheric wind abruptly and periodically changes direction from the west to an easterly phase and back again every 26 to 30 months; this shifting is known as the quasi-biennial oscillation.) The winter Arctic polar vortex is usually stronger and colder when the tropical high-altitude wind is blowing west, but at solar maximum the polar vortex remains weak and warm regardless of the wind direction.

This stratospheric effect might extend down through the atmosphere and around the globe, producing mid-winter warmings when solar activity is high and the wind is in its westerly phase. Indeed, Labitzke and Van Loon have shown that the troposphere grows hotter and cooler in step with the solar cycle in regions near the tropics. During periods of high solar activity, the increased radiative output from the Sun seems to intensify atmospheric circulation.

We do not know exactly how the solar variations are amplified to significantly influence our weather, and the exact physical mechanism for coupling the high-altitude wind variations to lower regions where the climate originates has not yet been found. So, while the correlations are impressive, the Sun-weather connection is just beginning to be understood and will remain controversial for a long time. Whatever the relationship, the existence of clear-cut connections between tiny variations in the Sun's radiative output and measurable changes on the Earth demonstrates that amazingly delicate balances are at work in the atmosphere.

the increase in global temperature caused by "unnatural" greenhouse warming over the same period is also 0.2 degrees. On time scales of a decade or less, the forcing of climate by low-level solar irradiance variability is now comparable to that of human-induced activities estimated by models of the greenhouse effect. It would therefore be unwise to exclude the solar modulation if we want to detect the effects of human activity in the global temperature record.

Of course, it is equally unwise to assume that the Sun is the main cause of climate change. Other natural effects, such as volcanic emissions, may noticeably affect global warming. Under the assumption that the increase in atmospheric greenhouse gases continues unabated, "unnatural" warming is expected to exceed that caused by the Sun by a factor of twenty within half a century, provided that the solar variations remain as small as those observed so far. Many scientists do not see how small changes in the Sun's energy output can create appreciable changes in the more energetic terrestrial weather system, so they assume that global warming will eventually outstrip any solar effect.

Yet, the inexorable increase in carbon dioxide may eventually be curtailed, and the solar changes may not always be relatively small and negligible. Current models indicate that a 2 percent decrease in the intensity of solar radiation would be required to cancel out the warming that is expected to result from doubling the atmospheric greenhouse gases relative to their pre-industrial levels. The doubling will occur by the mid-twenty-first century, if we keep on dumping the waste gases into the air at the present rate, but the Sun's variability could counteract the warming. (The climate may be cooled as well by small particles, called sulfate aerosols, emitted into the atmosphere by man-made industry and naturally occurring volcanoes – see Focus 10B.)

We cannot exclude the possibility of pronounced fluctuation in the solar output in the future, for the historical record and observations of other variable stars indicate that they might occur (see Sect. 10.6). So, scientists will want to monitor the Sun's radiative output in order to fully assess the future effects of human activity on our fragile environment.

In fact, variations in the Sun's luminosity might account for global surface temperatures observed during the past century. In 1991 the Danish scientists Eigil Friis-Christensen and Knud Lassen discovered a striking relationship between the length of the sunspot cycle and these temperatures. (The sunspot cycle has varied from 9.7 to 11.8 years since the mid-1800s, and its length has been compared to air temperatures in the Northern Hemisphere.) The two curves are closely intertwined (Fig. 10.3). In contrast, the global temperature record for the same period does not closely resemble the smooth increase calculated from the man-made greenhouse effect, leading some to suggest that the Sun may have played a role in global temperature change during the past century.

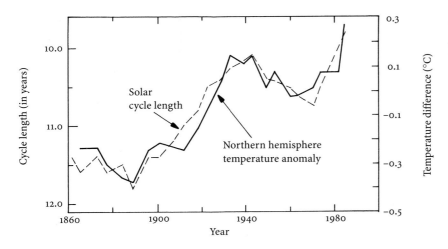

Fig. 10.3. Sunspot Cycle and Global Temperatures. Variations in the length of the sunspot cycle (*dashed line*) closely fit changes in the air temperature over land in the Northern Hemisphere (*solid line*). Shorter sunspot cycles seem to be associated with increased temperatures and more intense solar activity. This suggests that the Sun's radiation is at least partly responsible for the rise in global temperatures over the last century, and that the Sun can substantially moderate or enhance global warming brought about by increased carbon dioxide in the atmosphere. [Adapted from Eigil Friis-Christensen and Knud Lassen, Science 254, 698–702 (1991)]

That is not to say that "unnatural" greenhouse warming couldn't be important soon; its signature must eventually break through the temperature record if it hasn't already. If we continue to alter the composition of the atmosphere without control and indefinitely, our legacy will eventually include global warming. The continued accumulation of greenhouse gases in the air is inevitable, and the world may be getting hotter sometime soon, and quickly, even if we do not know exactly when. So, we must prepare for the full development of man-made global warming sometime in the twenty-first century.

10.4 CONSEQUENCES OF OVERHEATING THE EARTH

Once in the atmosphere, carbon dioxide stays there for centuries, so future generations will have to contend with the consequences of our present actions. The invisible waste gases that we have already dumped in the air will slowly change the climate of the Earth regardless of future actions.

The uncertainties in future forecasts of global warming are great, but a significant warming ought to occur sometime in the twenty-first century. When we reach that high-temperature stage, our life will certainly be changed. The potential danger lies not only in the world-wide temperature increase, but also in the rapid rate at which it might occur. Average global temperatures are likely to rise abruptly, in just a few decades, if humans keep putting invisible pollutants into the atmosphere at the present rate. Future global warming may take place so quickly and suddenly that natural ecosystems and human societies will not have time to adjust without great stress and potentially serious disruptions.

The predicted consequences of global warming range from the uncomfortable to the catastrophic, depending on how quickly it occurs and whether it occurs at the bottom or top of the predicted range (1.5 to 4.5 degrees Celsius when the greenhouse gases double). If it takes a gradual, modest course, some parts of the world could benefit, and the normal resilience of society ought to accommodate the climatic change. If the warming is rapid and severe, dire consequences could follow.

If the warming is in the lower range of the prediction, it could bring benefits. Increased rainfall would make some areas of the world more fertile, and crops and plants would be enriched and invigorated by rising concentrations of atmospheric carbon dioxide. The growing season would lengthen in the north, with longer summers and shorter winters, and residents of cities like Boston would have to shovel less snow and perhaps go skiing less often.

Even the worst scenarios of global warming, of nearly 5 degrees Celsius by the year 2030, might result in a climate in which living things could prosper, as they have in the remote past. One hundred million years ago, for example, it was 10 to 15 degrees Celsius warmer than today and all sorts of plants and animals flourished, including the dinosaurs. Even today, people experience extremes of temperature greater than that of any predicted global warming in the natural course of events. Seasonal temperature changes averaged over the Northern Hemisphere between winter and summer are 15 degrees Celsius. The average temperature differential between New York City and Atlanta, Georgia, or between Paris and Naples, is as large as the most extreme predictions of global warming, and there is little evidence of greater risk to people who live in the warmer southern climates. The capacity of humans to adapt is extraordinary, and most of us would probably welcome a little more heat or a move to a warmer place.

On the other hand, there is a long list of possible catastrophes, especially if there is rapid, significant warming at the high end of the predicted range. Some of the predicted environmental dangers are unprecedented in their seriousness and scope, at least throughout human history. The associated rise in sea level will flood coastal cities; tropical hurricanes will increase in intensity; water supplies will be reduced and forest fires will become more common; the air will become more polluted; drought will be intensified within the interiors of many continents; our bountiful farms will dry up; the American midwest will become a colossal dust bowl; power companies will be unable to air condition our sweltering cities; and extreme heat waves will cause great human stress and more deaths, particularly among the poor, elderly, and weak or those with cardiovascular and respiratory disease.

A rise in sea level is one of the most certain effects of the warming projected during the coming decades. Forecasters predict a half-meter rise in sea level as the result of global warming during the coming century. However, you won't notice the change, for the sea level is only rising a few millimeters every year.

As the water in the sea gets warmer, it will expand and ascend to higher levels at the edges of continents, in much the same way that heating the fluid in a thermometer causes the fluid to expand and rise up the thermometer's wall. This is because warm water or other fluids occupy a greater volume than cold ones. After allowing for expansion caused by the heating of the oceans, a secondary contribution to sea-level rise will be the melting of ice that now covers land, such as northern glaciers and ice sheets in Antarctica. But contrary to popular belief, the melting of ocean ice will not be the main cause of a rise in sea level. (When ice floating at sea, or in your drink at home, melts, it will not cause any change in water level, for it produces about the same volume that it displaces.)

A one-meter rise in sea level will seriously disrupt coastal areas where more than a quarter of the world's population now lives. Venice and Alexandria will be inundated, as will many cities on the Atlantic and Gulf coasts of the United States, including Boston and New York City. Flooding of the Ganges Delta in Bangladesh will create great human loss. The boundary between salt water and fresh water at the mouths of rivers will move several kilometers inland, so the Nile, Yangtse, Mekong, and Mississippi deltas are all at risk. Island nations will suffer severe coastal flooding or disappear altogether under the rising waters; they include Cyprus and Malta in the Mediterranean, many of the Caribbean islands and several Archipelagos around the Pacific Ocean.

Agricultural production and population centers in the United States and Europe would shift northward. Plants and animals would also migrate, as they have done throughout geological history, moving up and down in latitude as the globe warms or cools. But future global warming may be too rapid, and plants and forests may not be able to migrate to cooler climates fast enough, dying before they can take root and grow seeds.

Some studies indicate that about half the planet's vegetation could be disrupted. Many climate-sensitive habitats could be destroyed, hastening the extinction of some species, including the monarch butterfly, the white-flowered edelweiss, the Bengal tiger, the polar bear, and the walrus. Some claim that the very web of life is at risk, and that global climate change threatens the habitability of the planet.

10.5 LIFE IN THE FAST LANE

Humans, primarily those living in industrial societies, are tampering with the environment on a global scale. We now rival the grand natural forces like the movements of continents, volcanic eruptions, asteroid impacts, ice ages, and the varying Sun as an agent of global change. (A colliding asteroid is thought to have wiped out the dinosaurs and many other living things 65 million years ago.)

As the spiritual goes:

> He's got the whole world in his hands,
> the whole wide world in his hands.[50]

but now the "he" is not a supernatural deity – it's us. We have always exploited nature in the belief that the resilient fabric of our air, land, water, and life was so vast and enduring that people could never do it irrevocable harm. Now, as the result of rapid, recent, unprecedented population and industrial growth, nature is no longer independent of human activity and may not be able to sustain human growth without catastrophic effects. It is as if:

> Man is inescapable, everywhere on the globe, and nature is a
> fantasy, a dream of the past, long gone.[51]

Global warming as the result of human industrial activity seems inevitable sometime in the twenty-first century, and there is now an ongoing controversy about the threat. One side warns of an imminent doomsday of our own making, a human apocalypse caused by our tinkering with nature and changing the chemistry of the air. In their scenario, belching smokestacks, gasoline-powered automobiles, power-generating stations and the voracious destruction of forests are turning up the heat on an overburdened environment, pushing it over the edge. The sheer intensity of the peril has sparked the public's imagination. The poet Mark Strand reflected the anxiety:

> I dreamed that it was night. A thick debris
> Of clouds had magically dissolved, and we could see
> The ruins below – a couple cloaked in dust who used to lie
> All day in bed making love, but now, no more than bone
> And hair, are scattered on a floor of weathered stone ....
> It was years from now. A disembodied voice
> Was lecturing, "... . One wonders if they had a choice.
> Experts tell us all we know is autumn came,
> And the summer heat continued all the same.
> Others go further, saying the heat became a form of hell;
> Thermometers stuck, the oceans rose, an ashen curtain fell.
> When it dawned on everyone that things were wrong,
> The end was on its way, and nobody had long."[52]

The alternative view, equally cogent, is that the catastrophic predictions by apocalyptic pessimists are exaggerated and not firmly grounded on solid, unequivocal scientific evidence. Claims that the Earth is headed toward ecological disaster, with raging storms, flooded cities, and poisoned air are instead dismissed as wild ranting and speculations bordering on science fiction. Proponents of this perspective argue that the warmer temperatures and sharp climate swings of recent times may have nothing to do with a warming planet; they may be just normal fluctuations of the world's climate that we can't do anything about.

There are additional questions about the vast economic and social consequences that corrective action would involve. Staggering costs are foreseen if we are to curb the emission of carbon dioxide. A 20 percent reduction in the United States would run to trillions of dollars through the twenty-first century. The economic consequences of prevention may be greater than those of global warming if and when it occurs, so it might be cheaper to wait and deal with the effects of global warming rather than the possible causes. (On the other hand, German and Japanese industry have anticipated a vast market in environmentally benign technology.)

Some therefore say that we should wait and see if the forecasts can be trusted. Others argue that if we wait it will be too late, and the Earth and life on it will be irreversibly damaged.

Who should we believe? It's probably unwise to lapse into apocalyptic thinking, to promote public hysteria, and otherwise over-react, especially in view of the awesome scientific uncertainties. On the other hand, ostrich-like denial is also imprudent since the stakes are so high. Most scientists therefore support prudent steps to curb the continued build-up of heat-trapping gases despite the great scientific uncertainties about their current effects.

Many politicians are also concerned about the future; U. S. Senator Al Gore, and subsequently Vice President, reasoned:

> The total volume of all the air in the world is actually quite small
> compared to the enormity of the Earth, and we are filling it up,
> profoundly changing its makeup, every hour of every day,
> everywhere on Earth … . We are in fact conducting a massive,
> unprecedented – some say unethical – experiment. As we
> contemplate a choice between adapting to the changes we are
> causing and preventing those changes, we should bear in mind that
> our choice will bind not only ourselves but our grandchildren and
> their grandchildren as well … . What does it mean to redefine one's
> relationship to the sky? What will it do to our children's outlook on
> life if we have to teach them to be afraid to look up?[53]

Protective measures against the feared warming are nevertheless difficult to legislate, partly because the threat is remote and uncertain, unlikely to occur in our lifetimes, and also because most people don't like to be told what to do. Unlike the ozone hole, there is no clear proof that immediate action is required to halt global warming, and there is a concern that preventive legislation would unduly curtail economic growth and interfere with the free market. Oil companies are strenuously opposed to any limits on total carbon dioxide emission, and many individuals will resent higher electricity, heating, and transportation costs that might result from such limits.

The United States has the world's biggest economy and is the largest single emitter of carbon dioxide, both in total output and on a per capita basis. It is responsible for 18 percent of the world's carbon dioxide output, and a citizen of the United States uses ten to fifteen times as much energy as a citizen of India or China.

Yet, the government policy of the United States in the 1990s depends only on voluntary action by industry and individuals, with a goal of bringing the total national emission of carbon dioxide in the year 2000 back down to 1990 levels. There are no mandatory limits, no strict regulatory actions, and no corrective measures such as taxes for energy consumption. Thus far, this has been the only response to a distant crisis of uncertain magnitude, even when the country's Vice President has recommended stronger corrective measures.

Environmentalists would like the United States to take a tougher stance, and the leading economic power is sometimes portrayed as greedy, selfish, and irresponsible. For instance, in 1993 the Russian author Aleksander I. Solzhenitsyn wrote:

> When a conference of the alarmed peoples of the Earth convenes in the face of the unquestionable and imminent threat to the planet's environment, a mighty power, one consuming not much less than half of the Earth's currently available resources and emitting half of its pollution, insists, because of its own present-day interests, on lowering the demands of a sensible international agreement, as though it did not live on the same Earth. Then other leading countries shirk from fulfilling even these reduced demands. Thus, in an economic race, we are poisoning ourselves.[54]

Solzhenitsyn would like the industrialized countries to limit their desires and demands, to practice self-restraint and denial, to sacrifice for the future good of the world.

As previously noted, industrialized countries have declared their intention to limit emissions of carbon dioxide to keep the planet from overheating. Like the United States, the 12 countries of the European Union and Canada hope to reduce their total national emission of carbon dioxide to 1990 levels by the year 2000, but this informal statement of intent is not viewed as a legal obligation. Some of them, like Germany, have committed themselves to lower the levels even further.

How can we individually or collectively reduce the rate at which carbon dioxide is building up without jeopardizing economic growth? We can conserve energy and use it more efficiently. Houses and offices can be insulated so they require less heat, mass transit systems can be developed, and we can improve the efficiency of appliances, automobiles, lighting and other manufactured products. We can encourage the wider use of natural gas, since it produces less carbon dioxide than coal or oil, for the same energy production. Cars might run on batteries or solar power, and we can develop geothermal, hydroelectric, solar, wind, or other renewable sources of energy. And we can plant a lot more trees, which usually outlive humans, to extract more carbon dioxide from the air, while also preventing erosion and providing a home for diverse species.

However, the poor developing nations are not about to adopt restraints that might slow their industrial growth just to keep the rich, industrialized countries a little cooler. For instance, China has no plans to cut back on its carbon dioxide emissions, despite the fact that they will eventually have an unprecedented impact on the world's environment. (China's electric power plants, factories, and steel mills are among the least efficient in the world, and its reserves of soft coal dwarf those in the United States or anywhere else.) Forests in the developing countries are being cleared at a rapid rate, and their industries are often subsidized by practically giving away coal, oil, and natural gas in the hopes of improving the standard of living for their surging populations. If these global inequities are not corrected, the developing countries will contribute more to global warming than the richer ones by the year 2000.

People in the poorer nations argue that the average person in the rich countries eats more food, consumes more energy, and produces more than ten times more pollution than they do, and that wasteful over-consumption by the people in the rich countries, en route to their great prosperity, has contributed most of the environmental damage so far.

The poorer people of the world, mired in poverty and preoccupied with economic survival, are willing to help, but at a price. They will play the environmental game if the richer nations supply financial aid to help meet the cost of industrialization without damaging the environment. This gambit may be working, for money and technical assistance are being provided by the wealthier countries for cleaner cars and smokestacks in the poorer ones.

With these caveats in place, most of the world's governments signed a treaty in 1992, collectively agreeing to environmentally sustainable economic development that encourages economic growth without compromising the environment's ability to sustain it. In principle, this will safeguard our natural environment and assure the well-being of future generations, while the developing countries combat poverty and are integrated into the world's economy. Nevertheless, many of the poorer nations have not yet agreed to restrain population growth, halt the destruction of forests, or limit their emissions of heat-trapping gases, particularly carbon dioxide.

10.6 OVER THE LONG HAUL

Strong cosmic forces come into play when considering our fate over long intervals of thousands and millions of years. A hundred million years ago, when the dinosaurs roamed the Earth, the climate was some 15 degrees Celsius warmer everywhere than it is today, and there were no polar ice caps. Indeed, over most its 4.6 billion years the Earth has been free of ice, even at its poles. During this time, the Earth has been re-shaped by powerful internal forces, producing steady, irreversible changes in our global climate. Drifting continents collided to make towering mountain ranges, or split open to make way for new oceans, altering the flow of air or sea and paving the way for the growth of continental ice sheets.

During the last million years, the climate has been dominated by the recurrent, periodic ice ages. (The continents do not shift much in a million years, only about 10 kilometers.) During this stately cycle, great sheets of ice have rhythmically advanced and retreated several times. Roughly speaking, the climate has been one of glacial ice and cold punctuated every 100 000 years or so by brief intervals of unusual warmth, called interglacials, lasting about 10 000 years. We may now be approaching the end of an interglacial period.

At the height of the last ice age, about 5 percent of the sea water was frozen into glacial ice sheets. The oceans were some 100 meters lower than today, and it would have been possible to walk from England to France, from Siberia to Alaska, and from New Guinea to Australia. Glaciers advanced as far south as central Europe and the midwestern United States, burying everything in their path under ice a kilometer thick.

About 10 000 years ago, the present interglacial began – the one in which human civilization developed. The world then became warmer and wetter, about 5 degrees Celsius warmer on average. The ice sheets melted and shrank back to their present-day configurations, leaving only the glacial ice in Greenland and parts of arctic Canada, as well as the massive ice sheets of Antarctica. The sea level rose rapidly around the world, in a warm spell that has persisted until today, although interrupted from time to time by little ice ages.

Since no earlier interglacial has lasted more than 12 000 years, we could soon enter an ice age that might counteract any warming by human activity. Summer temperatures in the far northern latitudes are, in fact, no longer high enough to melt all the snow that falls in the winter, perhaps signaling the slow advance of the next ice age. Many thousands of years from now, Copenhagen, Detroit and Montreal could be buried under a mountain of ice instead of smoldering in the heat of global warming. However, the long transition to the next glacial period might occur gradually and slowly, while humans may be rapidly turning up the terrestrial thermostat. Still, during the next few centuries, the encroaching ice age could mask, offset or even surpass any effect caused by human interference with the environment.

The ponderous ebb and flow of the great continental glaciers is mainly due to long periodic changes in the global distribution of energy the Earth receives from the Sun. During its slow dance around the Sun, the planet moves in and out, nodding back and forth, and whirling and wobbling in a cosmic pirouette. This slowly alters the distances and angles at which sunlight strikes the Earth, thereby causing the ice ages.

This astronomical theory has been known, in various stages of development, for more than a century, but for most of that time it was not widely accepted. Joseph Alphonse Adhémar, a French mathematician, seems to have first suggested in 1842 that the ice ages might be due to variations in the way the Earth moves around the Sun. This idea was taken up in greater detail by James Croll, a self-taught Scotsman, who in 1864 showed how long periodic variations in the Earth's distance from the Sun might change the terrestrial climate. The theory received its fullest mathematical development in the 1930s and 1940s by Milutin Milankovitch, a Yugoslavian astronomer, who described how variations in the planet's wobble and tilt can influence the pattern of incoming radiation at different locations on the globe.

There are three astronomical rhythms, sometimes called the Milankovitch cycles, that apparently set the ice ages in motion by altering the seasonal distribution of the Sun's light and heat. The shortest of these is a periodic wobble in the Earth's rotation that is repeated mainly in periods of 23 000 years. It determines whether the seasons in a given hemisphere are enhanced or weakened by the orbital variations.

Scientists speculate that in periods when less sunlight falls in far northern latitudes, for example, less snow melts in the summer. Millennium after millennium, the snow is compressed into ice from which the glacial ice sheets grow and advance southward. Then, when more sunlight falls in the north, the ice sheets begin to melt.

The second of these astronomical rhythms is caused by the variation of the Earth's axial tilt from 21.5 degrees to 24.5 degrees and back again every 41 000 years. It is currently 23.5 degrees, and accounts for our yearly seasons. The greater the tilt is, the more intense the seasons in both hemispheres, with hotter summers and colder winters.

The third and longest cycle is due to a slow periodic change in the shape of the Earth's orbit every 100 000 years. Although the Earth's orbit is nearly circular, it is not perfectly so. It is slightly elliptical, or out of round. As the result of the shifting interplay of the gravitational tugs of the other planets, the orbit alternately elongates and contracts. This causes the Earth to move away from the Sun's heat and back again. As the orbit becomes more elongated, the Earth's distance from the Sun varies more during the year, intensifying the seasons in one hemisphere and moderating them in the other.

Fig. 10.4. Pacemakers of the Ice Ages. Temperatures recorded in the ancient fossilized shells of deep-sea sediment indicate that the major ice ages (white) during the past half million years recurred at intervals of 25 000, 40 000 and 100 000 years, with a dominant 100 000 year recurrence. These periodicities are caused by subtle changes in the Earth's orbit around the Sun and in the orientation of the Earth's rotational axis. They periodically change the amount and distribution of sunlight on the Earth, producing ice ages separated by warm intervals.

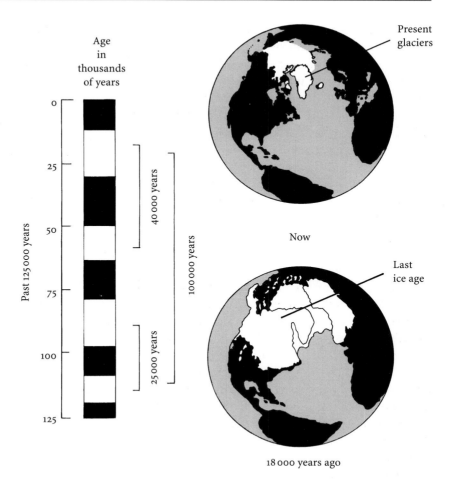

The astronomical theory for the recurring ice ages was not strongly supported until the 1970s when core samples of deep-sea sediments provided a clear signal of the three overlapping cycles (Fig. 10.4). The sediments are continually being laid down as tiny marine creatures, called plankton, die and fall to the bottom of the ocean. Chemicals in their chalky skeletons are used to reveal how much of the world's water was missing from the oceans, locked up in ice sheets when the plankton lived.

Scientists gauge the ice age/interglacial cycles from the relative amounts of two different types (isotopes) of oxygen atoms found in the deep-sea cores. There are eight protons in the nucleus of every oxygen atom, but two isotopes contain eight and ten neutrons, giving atomic mass numbers of 16 and 18, respectively. Molecules of water that contain the rare, heavier sort of oxygen, O_{18}, move more slowly than those of water with the more abundant, lighter type, O_{16}, so their relative amounts depend upon how cold it is.

Higher temperatures result in the evaporation of water vapor that contains relatively small amounts of the heavy isotope. When the tem-

perature falls, heavy water passes from the vapor to liquid stage more
readily than water made up of the lighter isotope. Continued cooling
gives rise to condensate with increasingly lower concentrations of the
heavy isotope. As the temperature decreases, you get relatively more of
the heavier oxygen in the calcium shells of sea animals. Likewise, during
glacial periods the removal of isotopically light water from the oceans to
form continental ice sheets leads to an increase in the ratio of heavy to
light atoms in the oceans as a whole. So, the ratio of the isotopes can be
used to infer the proportion of the world's water frozen within the glacial
ice sheets.

Investigations of sediment buried at the bottom of the sea has a fascinating history, told by John and Katherine Imbrie in their book *Ice Ages
– Solving the Mystery*. In capsule form, Cesare Emilani, a post-graduate
at the University of Chicago, showed in 1955 that oxygen isotopic variations in deep-sea cores reflect the ebb and flow of ice over the past
300 000 years. (Emilani thought that the isotope record acted as a thermometer, but in 1967 Nicholas Shackleton at Cambridge University
showed that the oxygen isotope data was not just a record of sea-sur-
face temperature, but a record of changes in the global ice volume.)
The 100 000-year pulsebeat of the climate was identified by Walter S.
Broecker and Jan van Donk, at Columbia University and its Lamont
Earth Observatory, after using magnetic data to establish the time scale
of variations in the oxygen-isotope ratios found in the sediment cores.
(The Earth's magnetic field has flipped, or reversed direction, at least 9
times over the past 3.6 million years; the timing of these reversals can be
established from the ages of rocks.)

Then a key paper on the subject, entitled "Variations in the Earth's
Orbit: Pacemaker of the Ice Ages" was published in *Science* on 10 December 1976 by Jim Hays of the Lamont-Doherty Geological Observatory,
John Imbrie from Brown University, and Nick Shackleton of the University of Cambridge. Their analysis of the isotopes in core sediments
showed that the major ice ages during the past half million years recurred at intervals of 23 000, 41 000 and 100 000 years, with a dominant
100 000 year recurrence. In other words, the isotope ratio and the global
amount of frozen water rose and fell in accordance with all three astronomical rhythms.

It was somewhat surprising that the glaciers have advanced and retreated every 100 000 years in synchronism with this rhythmic stretching of the Earth's orbit, and that this period dominates the deep-sea
data. The shorter cycles have a greater, direct effect on the seasonal
change in sunlight, but apparently produce smaller changes in ice volume than the longer one that has a weaker seasonal effect. By itself the
100 000-year cycle does not appear strong enough to bring about direct
alterations of the terrestrial climate, so it must be leveraged by some
other factor.

After all, the climate is shaped by the interaction of the Sun, land, sea, air and ice. They can act together to amplify effects triggered and timed by the astronomical rhythms. According to one theory, the varying distribution of sunlight, caused by the Milankovitch cycles, sets off changes in ocean circulation and the flow of hot and cold air. Indeed, redirected wind and weather patterns, that cooled the globe, seem to be responsible for the beginning of the ice ages, perhaps as the result of rising mountains such as the Tibetan Plateau and the Himalayas. Before that change the glaciers and recurrent ice ages did not exist, even though the cyclic variations in incident sunlight must have been present.

Nowadays, the Earth's past climate is deciphered from tubes of ice. Year after year, snow falls on the polar ice caps, the later snows slowly compressing the earlier ones into ice. Microscopic bubbles have been captured in the light and fluffy snow before it turned to ice, preserving pristine samples of the ancient atmospheric gases. Precise dates are obtained by counting the annual layers of ice, like the growth rings of a tree. The abundance ratio of heavy and light hydrogen atoms is used as a fossil thermometer, measuring the temperature at the time the snow fell. (Heavy hydrogen is known as deuterium, and the technique is similar to that using the two types of oxygen atoms.)

A comparison of the ancient air and temperatures was obtained from the Antarctica ice cores by French and Soviet scientists in 1987. The atmospheric carbon dioxide levels and global temperatures track each other unerringly, going up and down in tandem throughout the past 160 000 years (Fig. 10.5). The temperatures go up whenever the carbon

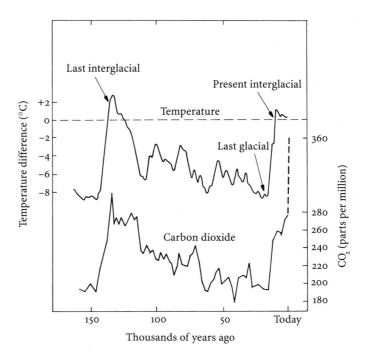

Fig. 10.5. Ice Ages and Carbon Dioxide. Changes in the Earth's temperature for the last 160 000 years closely parallel changes in the atmospheric carbon dioxide. However, we do not know if the fluctuations in carbon dioxide cause the temperature changes, or if they are both related to some other natural phenomenon. The temperatures fluctuate by up to 10 degrees Celsius in Antarctica, while the levels of carbon dioxide have varied by up to 100 parts per million. The ice core data does not include the past 200 years, shown as a dashed line on the right side of the carbon dioxide curve (also see Fig. 10.2). (Adapted from Claude Lorius, EOS, Vol. 69, No. 26, 1988)

dioxide does, so warm interglacial climates had more carbon dioxide in the air than the cool ice ages, by a factor of about one third. The carbon dioxide variations may amplify the slight changes in the Earth's temperature that are caused by the three astronomical cycles, producing more dramatic alterations in the temperature. (A comparison of Figs. 10.2 and 10.5 indicates that the concentration of carbon dioxide now in our air exceeds that at anytime in the past 160 000 years by another factor of one fourth, so future temperature increases might be imminent.)

However, we do not really know if the ancient carbon dioxide levels are a cause of the warming and cooling, for they could be an effect of the

FOCUS 10D

How Stable is the Earth's Climate

During the present interglacial, which began about 10 000 years ago, our planet has enjoyed a stable climate. It was therefore a surprise when ice cores suggested that the Earth's climate has changed relatively rapidly – both during times of glaciation and during interglacial periods. For instance, members of the Greenland Ice-Core Project, or GRIP, reported in 1994 that during the interglacial 130 000 years ago, when the Earth was as warm as it is today, there were very rapid climate changes characterized by abrupt warmings and severe cold snaps.

The ice-core data suggested that the current climate may become hot or cold much more quickly than had been believed, in sudden, violent shifts as brief as a decade or less rather than centuries. According to the climate experts, this increases the potential danger of disrupting the present climate with industrial waste gas.

The experts might have misinterpreted the data. A second, nearby deep-ice core agreed perfectly with the first one, except in the last interglacial period. When the ice in these deeper layers flows over bedrock, it churns up in waves that fold and mix the layers together. The scientists could have been misled by the ordered regularity of the upper ice layers, incorrectly assuming that it continued all the way down. So, our current climate may be quite stable after all.

Yet, the jury is still out on this matter, and informed scientists approach it in a circumspect manner. Recent studies of plant species, as traced by the pollen and spores deposited in European lake bottoms, suggest rapid climatic fluctuations during the last interglacial period, and huge swings in climate were routine during the last ice age. So, abrupt and frequent cycles of warmth and extreme cold are possible, and if applicable today heat-trapping gases like carbon dioxide might destabilize our climate. So, we have to consider the possibility.

temperature change. Indeed, scientists cannot yet decide whether the carbon dioxide fluctuations precede or follow the onset of the ice ages. Whatever the explanation, the methane concentration in the ice core bubbles also shows a close relation with climate during the past 160 000 years, doubling between cold glacial lows and warm interglacial highs, so biological processes may be involved.

Yet, other smaller, more frequent climate fluctuations are superimposed on the grand swings of the glacial/interglacial cycles (also see Focus 10D). These minor ice ages may result from variations in the activity of the Sun itself. Over time scales of several decades and centuries, sunlight apparently flickers and pulses like an old fluorescent light tube.

The changing weather has been documented in historical records going back thousands of years. We know, for instance, that in the North Atlantic region a warm climate between about 1100 and 1250, known as the Medieval Optimum, enabled Vikings to settle in Greenland and travel to the New World. Then, beginning in the mid-fifteenth century a 400-year chill caused the collapse of their colonies and prevented exploration of the North American continent.

Extremely cold spells, some lasting several decades, occurred in the northern hemisphere on several occasions over the past 500 years. Overall, the period is known as the "Little Ice Age", though there were warmer periods too. At the coldest times, mountain glaciers advanced, crops failed and villagers starved in parts of Europe (Fig. 10.6). The canals of Venice froze over during the severe cold, and Londoners were driving their carriages across the frozen Thames. It was the time of the fictional Hans Brinker of the Silver Skates who raced on the frozen Dutch canals.

Fig. 10.6. Hunters in the Snow, 1565. During the prolonged period between 1500 and 1850, the average temperatures in Northern Europe were much colder than they are today. This chilly spell, known as the Little Ice Age, lives on in this painting by Pieter Bruegel the Elder. The coldest part of this period coincides with a conspicuous absence of sunspots and other signs of solar activity, called the Maunder Minimum (see Fig. 10.7). This suggests a correlation between Earth's climate and the Sun's behavior, with cold weather related to a lapse in activity on the Sun. (Photo by P. Meyer, Courtesy of the Kunsthistorisches Museum, Vienna)

Between 1645 and 1715, sunspots virtually disappeared from the face of the Sun. The German astronomer Gustav Spörer called attention to the 70-year absence in 1887. The missing sunspots were fully documented a few years later by E. Walter Maunder at the Royal Greenwich Observatory in England, so this period is now known as the Maunder Minimum. Its effects were originally reported in 1733 by the French author Jean Jackues D'Ortous de Mairan, as a decrease in the number of auroras seen on Earth, but he was ridiculed for thinking that the northern lights could be related to increases in the number of sunspots.

The importance of the absence of sunspots lies not in the sunspots themselves, but in the associated reduction of solar activity. Recent small variations in the Sun's total radiative output track the ups and downs of the solar magnetic activity cycle (see Sect. 9.5). It is therefore suspected that larger fluctuations in the total amount of sunlight are signaled by the presence or absence of sunspots. Thus, the dearth of sunspots during the Maunder Minimum is thought to be related to a marked decrease in the Sun's radiative output, producing unusually cold weather. (Although it was cold in many areas during the Maunder Minimum, there were colder episodes during the longer "Little Ice Age".)

The wood of very old trees tells us that the Maunder Minimum was not an isolated phenomenon. Trees reflect changes in the weather through alterations in their annual growth rings. The ages of the rings are determined by counting back from the present at the rate of one ring per year. Moreover, solar activity can be estimated from the amount of radioactive carbon in the tree rings of old, but still living, trees.

FOCUS 10E
Varying Stars

Young stars, like children, are more active, and more impulsive and irregular in their behavior than older stars like the Sun. They also have exceptionally large dark spots on their surface, and tend to vary more in total luminosity than the Sun does. The youngest stars generally rotate faster than older ones, and presumably generate stronger magnetic fields; they therefore are more intense and variable in invisible ultraviolet and X-ray radiation.

Younger stars also tend to become fainter as their magnetic activity level increases. In contrast, the Sun and other older stars become brighter at times of increased activity. It appears that during its lifetime, the Sun has progressed from a stage when the luminosity variation over its activity cycle was dominated by dark spots, to the present situation when it is dominated by bright faculae.

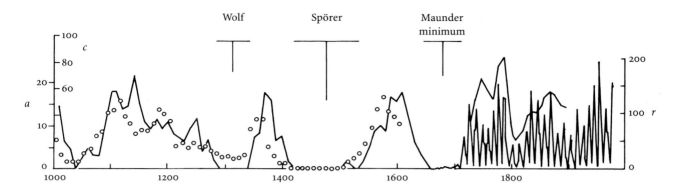

Fig. 10.7. Lapse in Solar Activity. Three independent indices confirm the existence of long-period changes in the level of solar activity. The observed annual mean sunspot numbers (*r*, scale at right) is shown from AD 1650 to AD 1990; the sunspot number was so low between 1650 and 1715 that it is not possible to trace the 11-year cycle of solar activity that becomes pronounced after about AD 1700. The English astronomer, E. Walter Maunder, first documented the missing sunspots with historical records, hence the name Maunder Minimum. The curve extending from AD 1000 to AD 1900 is a proxy sunspot number index derived from measurements of carbon 14 in tree rings (*c*, scale at *left*). Open circles are an index of the occurrence of northern hemisphere auroras (*a*, scale at *left*), another measure of solar activity, in sightings per decade. The tree ring data substantiate the Maunder Minimum, while both the auroral and tree ring results indicate two other pronounced minima in solar activity, labeled Wolf and Spörer. (Adapted from data provided by John A. Eddy, University Corporation for Atmospheric Research)

Cosmic rays from outer space can produce radioactive carbon,[14]C, when they strike atoms in the air. Because the cosmic rays are deflected away from the Earth by the Sun's magnetic fields during high solar-activity levels, there is less radioactive carbon in the air during episodes of high solar activity, and more of it at times of low activity on the Sun. Because radioactive carbon decays into normal lighter carbon,[12]C, at a known rate, the ratio of the two types of atoms in a tree ring analyzed today reveals how much radioactive carbon was present in the atmosphere when the ring was formed. This also tells us how active the Sun was at the time that the radioactive carbon was assimilated.

Such an analysis of the world's longest-lived trees, the bristle cone pines, suggests that the Sun's output has been turned down for several extended periods in the past millennia. John Eddy used the technique to read the history of solar activity all the way back to the Bronze Age, and showed that the tree-ring data are supported by other evidence such as the ancient sightings of terrestrial auroras (Fig. 10.7). In an important paper entitled "The Maunder Minimum", published in *Science* on 18 June 1976, Eddy concluded that the Sun has spent nearly a third of the past two thousand years in a relatively inactive state. He pinpointed several periods of low activity, each about a century long, naming them the Maunder, Spörer and Wolf minima (see Fig. 10.7).

The prolonged periods of low solar activity, and increased radioactive carbon, correspond closely to cold spells in the climate record. Indeed, there is now clear corroborating evidence from radioactive carbon in tree rings, as well as radioactive beryllium in layered ice cores, that several periods of reduced activity on the Sun, each of 40 to 100 years duration, have occurred every two centuries for the past millennia, and that these minima appear to have coincided with colder-than-average temperatures.

By how much does the Sun have to vary to produce a modest ice age? Temperatures in northern Europe may have dipped as much as 1.0 degrees Celsius during the coldest periods of the "Little Ice Age" which coincided with decreased solar activity. Current models indicate that a decrease in the Sun's radiative output of between 0.2 and 0.5 percent over

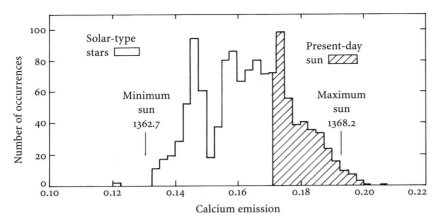

several decades could explain this cooling. Although this is somewhat larger than the 0.1 percent solar-cycle variation that has been measured so far, we have not yet entered a comparable lull in solar activity and there are still plenty of sunspots on the Sun.

Recent observations of other similar stars have shown substantial variations in stellar radiation (see Focus 10E). Solar-type stars, with the same mass, composition and age as the Sun, often undergo drastic changes of 0.5 percent or so in luminosity. Their chromospheric emission shows cyclic variations over periods similar to the Sun's magnetic-activity cycle. And since roughly one-third of the stars monitored exhibited no activity cycle, prolonged intervals of inactivity may not be uncommon. Thus, observations of Sun-like variable stars indicate that small, persistent variations in the solar energy output could produce extended periods of global cooling such as those associated with the Little Ice Age and the Maunder Minimum (Fig. 10.8).

Humans will continue pumping up atmospheric carbon dioxide over the next several decades, and the resultant global warming cannot be offset by a comparable lull in solar activity. That would require a more substantial decrease of about 1 or 2 percent in the Sun's luminosity. (Alternatively, a major ice age of the sort driven by the astronomical rhythms might do the trick.) So, we really don't know if our fate is one of fire or ice. I would side with Robert Frost who wrote:

Some say the world will end in fire,
Some say in ice.
From what I've tasted of desire
I hold with those who favor fire.
But if it had to perish twice,
I think I know enough of hate
To say that for destruction ice
Is also great
And would suffice. [55]

Fig. 10.8. Magnetic Activity in Solar-Type Stars. The strength of calcium emission (in the *H* and *K* Fraunhofer lines) provides an indication of magnetic activity on the present-day Sun (*hatched area*) and solar-type stars (*solid line*). The broad distribution at higher values of magnetic activity (*right*) corresponds to variable stars. The narrower peak (*left*) is attributed to currently invariable stars, and may represent stars in a Maunder Minimum. The total irradiance detected during the most recent maximum in the solar-activity cycle (1368.2 thousand erg/s/square centimeter) is, for example, compared with that expected during a Maunder Minimum (1362.7). Although the range in activity of the present-day Sun is typical of about one third of the solar-type active stars, the large number of such stars in a lower state of activity suggests that the Sun may undergo larger fluctuations in output associated with dramatic changes in the Earth's climate. [Adapted from Sallie Baliunas and Robert Jastrow, Nature *348*, 520–523 (1990) and Judith Lean et al., Geophysical Research Letters *19*, 2203–2206 (1992)]

Glossary

Absorption The process by which the intensity of radiation decreases as it passes through a material medium. The energy lost by the radiation is transferred to the medium.

Absorption line A dark line at a particular wavelength of the electromagnetic spectrum, formed when a cool, tenuous gas, between a hot, radiating source and an observer, absorbs electromagnetic radiation of that wavelength. These features look like a line when the radiation intensity is displayed as a function of wavelength; such a display is called a spectrum. Different substances produce characteristic patterns of absorption lines.

Acoustic waves Sound waves that propagate through the gaseous layers of the Sun's interior. They are produced when pressure is the dominant restoring force and there is a small displacement of the gas. Sound waves in the Sun produce oscillations of its visible surface with periods of about five minutes; observations of these oscillations can be used to infer the interior constitution of the Sun. *See* five minute oscillations and helioseismology.

Active region A region in the outer layers of the Sun where solar activity is taking place. The magnetized realm in, around and above sunspots is a disturbed area called an active region. It develops when strong magnetic fields emerge from inside the Sun. Radiation from active regions is enhanced over the whole electromagnetic spectrum, from X-rays to radio waves, though in sunspots the localized decrease in temperature reduces the visual brightness. Active regions may last from several hours up to a few months, and their number varies in step with the 11-year sunspot cycle. *See* solar activity cycle and sunspot cycle.

Activity index Any one of a number of indicators of the level of solar activity at a given time. Indices to measure solar activity include the number of sunspots, the total area covered by sunspots on the visible hemisphere, the areas and brightness of plages, and the total radio and X-ray radiation from the Sun.

Alpha particle The nucleus of a helium atom, consisting of two protons and two neutrons.

Arc degree A unit of angular measure of which there are 360 in a full circle.

Arc minute A unit of angular measure of which there are 60 in 1 arc degree.

Arc second A unit of angular measure of which there are 60 in 1 arc minute and therefore 3600 in 1 arc degree.

Astronomical unit The average distance between the Earth and the Sun, about 150 million kilometers.

Atmosphere The outermost gaseous layers of a planet, natural satellite or star. Since gas has a natural tendency to expand into space, only bodies that have a sufficiently strong gravitational pull can retain atmospheres. We use the term atmosphere for the tenuous outer material of the Sun because it is relatively transparent at visible wavelengths. The solar atmosphere includes, from the deepest layers outward, the photosphere, the chromosphere, and the corona. The Earth's atmosphere is transparent and consists mainly of molecular nitrogen (78 percent) and molecular oxygen (21 percent) with trace amounts of carbon dioxide (0.36 percent).

Atmospheric window A wavelength band in the electromagnetic spectrum that is able to pass through the Earth's atmosphere with relatively little attenuation through absorption, scattering or reflection. There are two main windows – the optical window and the radio window.

Aurora A display of rapidly varying, colored light given off by collisions between charged particles trapped in a planet's magnetic field and atoms of atmospheric gases near the planet's magnetic poles. Auroras are visible on Earth as the aurora borealis, or northern lights, and the aurora australis, or southern lights. They include the green and red emission lines from oxygen atoms and nitrogen molecules that have been excited by electrons from the Sun. Auroras occur at heights of around 100 kilometers.

Auroral oval One of the two oval-shaped zones on the Earth in which the nighttime auroras are observed most often. They are located at latitudes of about 67 degrees north and south, and are about 6 degrees wide. Their position and width both vary with geomagnetic activity.

Beta decay Emission of an electron and a neutrino by a radioactive nucleus.

Beta particle An electron or positron (i.e. a particle with the same mass as an electron but opposite electric charge) emitted from an atomic nucleus as a result of a nuclear reaction or in the course of radioactive decay. Such particles were originally called beta rays.

Black body A body that absorbs all the radiation incident on it. The intensity of the radiation emitted by a black body and the way it varies with wavelength depend only on the temperature of the body and can be predicted by quantum theory.

Black body radiation The radiation emitted by a black body. The wavelength of the peak emission is inversely proportional to the surface temperature. The total amount of energy emitted by a black body is proportional to the product of its surface area and the fourth power of its surface temperature. *See* thermal radiation.

Blueshift The Doppler shift in wavelength of spectral lines towards shorter wavelengths. Blueshifts arise when the source of radiation and its observer are moving towards each other. *See* Doppler shift.

Bow shock The boundary on a planet's sunlit side where the solar wind is deflected and there is a sharp decrease in its velocity. The plasma of the solar wind is heated and compressed at the bow shock.

Bremsstrahlung Radiation emitted when a free electron is accelerated in the vicinity of an ion but remains in a hyperbolic orbit without being captured. It is also called a free-free transition because the electron is free both before and after the encounter.

Burst Suddenly enhanced nonthermal radio emission from the solar corona during a solar flare, caused by energetic electrons in the coronal magnetic field. Bursts are divided into several types, depending on their time-frequency behavior.

Butterfly diagram A representation in graphical form of the way the latitudes at which sunspots appear vary as a function of time throughout the solar cycle. It was first plotted in 1922 by E. Walter Maunder and is also known as the Maunder diagram. The plot resembles butterfly wings, which gives the diagram its popular name.

Charged particles Fundamental components of subatomic matter, such as protons and electrons, that have electrical charge.

Chromosphere The part of the solar atmosphere between the photosphere and the corona. The Sun's temperature rises to about 10 000 degrees Kelvin in the chromosphere. The name literally means "sphere of color"; the chromosphere is seen as a pinkish glow during a total solar eclipse. The chromosphere has an emission spectrum and is routinely observed in the red light of hydrogen alpha. Spicules containing chromospheric material penetrate well into the corona (to heights of 10 000 kilometers) at the edges of the supergranulation cells. *See* hydrogen alpha and spicules.

Chromospheric network A large-scale cellular pattern visible in spectroheliograms taken in hydrogen alpha and other spectral regions. The network appears at the boundaries of the photospheric supergranulation cells, and contains magnetic fields that have been swept to the edges of the cells by the flow of material in the cell.

Convection The physical upwelling of hot matter, thus transporting energy from a lower, hotter region to a higher, cooler region. A bubble of gas that is hotter than its surroundings expands and rises. When it has cooled by passing on its extra heat to its surroundings, the bubble sinks again. Convection can occur when there is a substantial decrease in temperature with height such as the Sun's convection zone.

Convection zone A layer in a star in which convection currents are the main mechanism by which energy is transported outward. In the Sun, a convection zone extends from just below the photosphere to about seventy percent of the solar radius.

Core In solar astronomy, the heart of the Sun, where energy is generated by nuclear reactions. *See* Fusion.

Corona The outermost atmosphere of the Sun consisting of highly rarefied gas heated to temperatures of millions of degrees. The solar corona expands to form the solar wind. The corona is visible to the naked eye during a total eclipse of the Sun. The corona can always be observed across the solar disk at X-ray and radio wavelengths. *See* F corona and K corona.

Coronagraph An instrument for observing the Sun's corona in white light, which is normally seen only at a total solar eclipse. Invented by Bernard Lyot in 1930, the coronagraph is a special telescope in which an occulting, or blocking, disk produces an artificial eclipse of the Sun, enabling the instrument to photograph the faint light of the corona. However, even with a coronagraph located at a site where the sky is very clear, scattering of light by the Earth's atmosphere is a problem. Coronagraphs therefore work best when placed on satellites above the Earth's atmosphere.

Coronal green line An emission line of Fe XIV at 5303 Angstroms, the strongest visible line in the solar corona.

Coronal hole An extended, low-density, low-temperature region in the solar corona where the extreme ultraviolet and X-ray coronal emission is abnormally low or absent. The low density of the gas makes this part of the corona appear dark in X-ray images of the Sun, as if there were a hole in the corona. They typically last for several rotations of the Sun, and are always present near the solar poles. Particles from the Sun stream along the open magnetic fields in coronal holes to form the high-speed part of the solar wind.

Coronal mass ejection A vast magnetic bubble of plasma that erupts from the Sun's corona and travels through space at high speed. Coronal mass ejections may cause intense geomagnetic storms and accelerate vast quantities of energetic particles. *See* geomagnetic storm.

Coronium A supposedly unknown chemical element emitting unidentified emission lines in the spectrum of the solar corona. It was later discovered that the mysterious lines are produced by highly ionized forms of known elements, such as iron.

Corpuscular radiation Charged particles (mainly protons and electrons) emitted by the Sun. The stream of electrically charged particles was hypothesized in the 1950s and later renamed the solar wind. *See* solar wind.

Cosmic rays High-energy charged particles travelling through interstellar space at nearly the velocity of light. Those beyond the Earth's atmosphere are known collectively as primary cosmic rays.

Cyclotron radiation Electromagnetic radiation emitted by electrons travelling in circular paths in a magnetic field.

D layer The lowest part of the Earth's ionosphere at heights between about 50 and 90 kilometers. This is the layer that reflects radio waves.

Differential rotation The rotation of a gaseous body, such as the Sun, at a rate that varies with latitude. A solid body like the Earth must rotate so that the angular velocity is the same everywhere, but the equatorial regions of a gaseous planet or star rotate more quickly than those at higher latitudes. Thus, the outer part of the Sun revolves faster at its equator than at its poles.

Doppler effect The change in the observed frequency of sound or electromagnetic radiation when the source of waves and the observer are moving toward or away from each other.

Doppler shift A change in the wavelength or frequency of the radiation received from a source due to its relative motion along the line of sight. A Doppler shift in the spectrum of an astronomical object is commonly described as a redshift when it is towards longer wavelengths (object receding) and as a blueshift when it is towards shorter wavelengths (object approaching). The Doppler shift makes it possible to determine the radial velocity. *See* blueshift, radial velocity and redshift.

Eclipse The obscuration of light from a celestial body as it passes through the shadow of another body. In a solar eclipse, the Moon passes between the Sun and the Earth, blocking (or occulting) the Sun. A solar eclipse can occur only at or very close to new Moon, and it has a maximum duration of $7\frac{1}{2}$ minutes. During the brief moments of a total solar eclipse, darkness falls, and the outer parts of the Sun, the chromosphere and the corona, are seen. At any given point of Earth's surface, a total solar eclipse occurs, on the average, once every 360 years.

Ecliptic Plane of the Earth's orbit.

Electromagnetic radiation Radiation that travels through vacuous space at the speed of light and propagates by the interplay of oscillating electrical and magnetic fields. Electromagnetic radiation, in common with any wave, has a wavelength and a frequency; their product is equal to the velocity of light, so the wavelength decreases when the frequency increases. The energy associated with the radiation increases in direct proportion to its frequency.

Electromagnetic spectrum The entire range of all the various kinds or wavelengths of radiation including gamma rays, X-rays, ultraviolet radiation, and radio waves; light (or the visible spectrum) comprises just one small segment of this much broader spectrum. *See* spectrum.

Electron A negatively charged elementary particle that resides outside (but is bound to) the nucleus of an atom; free electrons have broken free of their atomic bonds and are not bound to atoms.

Electron volt (symbol eV) A unit of energy, principally used in atomic and molecular physics. It is defined as the energy acquired by an electron when it is accelerated through a potential difference of 1 volt in a vacuum. $1\text{ eV} = 1.60 \times 10^{-12}$ ergs. Electron volts are convenient units for measuring the energies of particles and electromagnetic radiation. The energies of X-rays are expressed in thousands of electron volts (keV). Millions (MeV) and billions (GeV) of electron volts are also used as units for the highly energetic atomic particles. An electron with a kinetic energy of a few MeV is travelling at almost the velocity of light.

Emission line A bright line at a particular wavelength of a spectrum, emitted directly by a hot gas, and revealing by its wavelength a chemical constituent of that gas.

Erg The cgs unit of energy; the work done by a force of 1 dyne acting over a distance of 1 centimeter.

Evershed effect The radial flow of gas in the penumbra of a sunspot, discovered by John Evershed in 1908.

F corona One of the components of the solar corona caused by light scattered from dust particles in the vicinity of the Sun. The "F" stands for Fraunhofer; the spectrum of the F corona is that of the Sun, including Fraunhofer lines.

Faculae Bright regions of the photosphere seen in white light, visible only near the limb of the Sun. They are brighter than the surrounding medium due to their higher temperatures and greater densities.

Filament A dense, massive structure that lies above the chromosphere, generally along a line separating regions of opposite magnetic polarity. Filaments appear as long dark features when observed on the solar disk in the light of certain spectral lines, particularly hydrogen alpha and those of ionized calcium. A filament is a prominence seen in projection against the solar disk. *See* prominence.

Fission A nuclear process that releases energy when heavy nuclei break down into light ones.

Five-minute oscillations Vertical oscillations of the solar atmosphere with a well-defined period of five minutes, usually interpreted in terms of trapped sound waves.

Flare Rapid release of energy from a localized region on the Sun in the form of electromagnetic radiation and energetic particles. Flares usually last for a few minutes, during which matter can reach temperatures of hundreds of millions of degrees. Most of the radiation is emitted as X-rays but flares are also observed at hydrogen alpha and radio wavelengths. They are associated with active regions of the Sun and are probably caused by the sudden release of large amounts (about 10^{32} ergs) of magnetic energy in a relatively small volume in the solar corona. *See* burst.

Flocculi A fine mottled pattern seen on the Sun when the solar chromosphere is imaged in monochromatic light, such as the light of singly ionized calcium.

Footpoint Intersection of magnetic loops with the photosphere.

Forbidden lines Spectral lines not normally observed under laboratory conditions because they have an intrinsically low probability of occurrence, or because they result from a transition between a metastable excited state and the ground state. Under typical conditions, an atom in a metastable state will lose energy through a collision before it is able to decay to the ground state. Under astrophysical conditions, however, where there are huge numbers of atoms and densities may be very low, it is possible for such "forbidden" transitions to occur and to produce strong spectral lines.

Forbush decrease A decrease in cosmic-ray intensity with an increase in solar activity (and *vice versa*). This phenomenon was first noted by Scott Forbush in 1954.

Fraunhofer lines The dark absorption lines in the spectrum of the Sun. Many of the stronger ones were first specified by Josef von Fraunhofer in 1814, who also labelled some of the most prominent with letters of the alphabet. Some of these identifying letters are still commonly used, notably the sodium D lines and the calcium H and K lines.

Frequency The number of crests of a wave passing a fixed point each second usually measured in units of Hertz (1 Hz = one oscillation per second).

Fusion A nuclear process that releases energy when light nuclei combine to form heavy ones. Fusion powers hydrogen bombs and makes stars shine.

Gamma rays The most energetic form of electromagnetic radiation, with the highest frequency and the shortest wavelength. Because Earth's atmosphere absorbs the radiation at that end of the spectrum, gamma ray studies of the Sun are conducted from space.

Geomagnetic field The magnetic field in the vicinity of the Earth. To a first approximation, the Earth's magnetic field is like that of a bar magnet (dipole) currently displaced about 500 kilometers from the center of the Earth towards the Pacific Ocean and tilted at 11 degrees to the rotation axis.

Geomagnetic storm A rapid change, typically of a few hours' duration, in the Earth's magnetic field caused by the arrival in the vicinity of the Earth of eruptions from the Sun. Auroral activity and disruption of radio communications are common during such storms.

Geosynchronous orbit An orbit around the Earth at an altitude where a satellite moves at just the speed at which the planet rotates; hence, an orbit in which an orbiting satellite remains nearly stationary above a particular point on the planet.

Granulation A cellular pattern observed in high-resolution images of the Sun's photosphere, caused by the convective movement of hot gases rising from hotter layers at greater depths.

Granule A roughly circular region on the Sun whose bright center indicates hot gases rising to the surface and whose dark edges indicate cooled gases that are descending toward the interior. Individual granules appear and disappear on time scales of about five minutes and exhibit a range of sizes, with 1000 kilometers being a typical diameter.

Greenhouse effect The trapping of infrared heat radiation by an atmosphere, thus causing greater surface heating than would normally be the case if the infrared radiation could escape directly into space. The natural greenhouse effect warms the surface of the Earth by about 31 degrees Celsius and keeps the oceans from freezing. How much heating the greenhouse effect causes depends on how opaque the atmosphere is to infrared radiation. Water vapor and carbon dioxide are the main sources of the opacity, but other trace gases like methane also play a role. Concern has mounted that global warming of the Earth will result from increased concentrations of carbon dioxide and other so-called "unnatural" greenhouse gases released by human activity, particularly the burning of fossil fuels such as coal and oil.

H and K lines The strongest lines in the visible spectrum of ionized calcium, lying in the violet at wavelengths of 3934 and 3968 Angstroms. They are conspicuous features in the spectra of many stars, including the Sun. The designations H and K were given by Fraunhofer and are still commonly used. *See* Fraunhofer lines.

Helioseismology The study of the interior of the Sun by the analysis of its natural modes of oscillation, which are observed spectroscopically as Doppler shifts in the absorption line spectrum. It includes investigations of sound waves trapped in the Sun, and provides information about the solar interior in much the same way that seismic studies of the Earth reveal details of its internal structure. Helioseismology is a hybrid name combining the Greek words *helios* for the Sun and *seismos* for earthquake.

Heliosphere The region around the Sun where the solar wind dominates the interstellar medium.

Heliostat A moveable flat mirror used to reflect sunlight into a fixed solar telescope.

Helmet streamer Named after spiked helmets once common in Europe, helmet streamers form over the magnetic neutral lines in large active regions with a long-lived prominence commonly embedded in the base of the streamer. Higher up, the magnetic field is drawn out into interplanetary space. Gas flowing out along these open magnetic fields might help to create the slow-speed solar wind.

Hydrogen alpha Light emitted with a wavelength of 6563 Angstroms from an atomic transition in hydrogen, the lowest energy transition in its Balmer series. This wavelength is in the red portion of the visible spectrum and is the dominant emission from the solar chromosphere.

Hydrostatic equilibrium The condition of stability that exists when inward gravitational forces are exactly balanced by outward gas and radiation pressure.

Ice Age A period of cool, dry climate causing a long-term buildup of ice far from the poles. The ice ages are caused by changes in the amout and distribution of sunlight at the Earth.

Interplanetary medium The medium between the planets in the solar system composed of interplanetary dust, electrically charged particles from the Sun and neutral gas from the interstellar medium. The charged particles consist of electrons, protons and helium nuclei (alpha particles) streaming outwards from the Sun and forming the solar wind.

Interplanetary scintillation Fluctuations in the signal received from a distant radio source observed along a line of sight close to the Sun. The scintillation is caused by irregularities in the solar wind.

Invisible radiation Those kinds of radiation to which the human eye is not sensitive; for example, radio and ultraviolet waves, as well as X-rays and gamma rays.

Ion An atom that has lost or gained one or more electrons, thus becoming electrically charged. By contrast, a neutral atom has an equal number of negatively charged electrons and positively charged protons, giving the atom a zero net electrical charge.

Ionization The process by which ions are produced, typically by collisions with atoms or electrons, or by interaction with electromagnetic radiation.

Ionosphere An ionized layer in the Earth's atmosphere that lies between heights of about 50 and 300 kilometers, though the extent varies considerably with time, season and solar activity. It is created by ultraviolet and X-ray radiation from the Sun. The D region, between 50 and 90 kilometers, has low electron density. The E and F regions, at about 100 and 200 to 300 kilometers, form the main part of the ionosphere.

Isotope An atom with the same number of protons as the most common form of the element but a differing number of neutrons.

K corona The inner part of the solar corona which emits a continuous spectrum without absorption lines. The "K" comes from the German Kontinuum. Physically, the K component results from Thomson scattering of photospheric radiation by free electrons in the corona. The K component is polarized and decreases rapidly in intensity with distance from the Sun.

Kilometer A unit of distance equal to 0.6214 mile.

Kinetic energy The energy of an object due to its motion.

Light The kind of radiation to which the human eye is sensitive.

Limb The apparent edge of the Sun (or Moon) as it is seen in the sky. Astronomers refer to the left edge of the solar disk as the Sun's east limb and to the right edge as its west limb.

Magnetic field A field of force around the Sun and the planets, generated by electrical currents. The Sun's magnetic field, like that of Earth, exhibits a north and south pole linked by lines of magnetic force.

Magnetic storm *See* geomagnetic storm.

Magnetograph An instrument used to map the strength, direction and distribution of the magnetic field across the surface of the Sun.

Magnetopause The boundary layer between the magnetosphere and the solar wind.

Magnetosphere The region surrounding the Earth, or any other planet, in which the planet's magnetic field has a controlling influence on the motions of charged particles. The magnetosphere is shaped by interactions between a planet's magnetic field and the solar wind.

Magnetotail A comet-like extension of a planet's magnetosphere formed on the planet's dark night side by the action of the solar wind.

Maunder diagram *See* butterfly diagram.

Maunder minimum The period roughly between 1645 and 1715 when there was an apparent lack of sunspots and auroras. Named after E. Walter Maunder.

Minute of arc *See* arc minute.

Molecule A tightly knit group of two or more atoms, bound together by electromagnetic forces among the electrons and nuclei of the atoms.

NASA Abbreviation for National Aeronautics and Space Administration.

National Aeronautics and Space Administration (NASA) The United States government agency responsible for civilian manned and robotic activities in space, including launch vehicles, scientific satellites, and space probes.

Neutral region A region where the longitudinal magnetic field strength approaches zero. Generally, neutral regions occur between regions of opposite polarity.

Neutrino An elementary particle with no electric charge and almost no mass, which interacts very weakly with other matter. It travels at the velocity of light and is produced in vast quantities by the nuclear reactions that take place in the center of the Sun.

Neutrino astronomy The attempt to detect neutrinos from cosmic sources, especially the Sun. Because they hardly interact with matter at all, neutrinos are very difficult to detect.

Neutron A neutral (no electric charge) elementary particle having slightly more mass than a proton, and which resides in the nucleus of most atoms.

Nonthermal radiation The electromagnetic radiation produced by an electron travelling at a speed close to that of light in the presence of a magnetic field. Such radiation is called synchrotron radiation after the man-made particle accelerator where it was first seen. More generally, the term nonthermal radiation is used for any electromagnetic radiation not produced by thermal processes.

Northern lights A popular name for auroras when observed from northern latitudes.

Nuclear force The force that binds protons and neutrons within atomic nuclei, and which is effective only at distances less than 10^{-13} centimeter.

Nucleus The positively charged core of an atom, comprising protons and (except for hydrogen) neutrons, around which electrons orbit.

Opacity A measure of the ability of a gaseous atmosphere to absorb radiation and become opaque to it. A transparent gas has little or no opacity.

Ozone A form of molecular oxygen containing three atoms (O_3) instead of the normal two. It is created by the action of ultraviolet sunlight on the Earth's atmosphere.

Ozone layer A region in the lower part of Earth's stratosphere (about 20–60 kilometers above sea level) where the greatest concentration of ozone (O_3) appears. The ozone layer shields the Earth's surface from the Sun's lethal ultraviolet rays.

Pair annihilation Mutual annihilation of an electron and positron with the formation of radiation (511 keV gamma rays).

Penumbra The lighter periphery of a sunspot, surrounding the darker umbra.

Photoionization The ionization of an atom by the absorption of a photon of electromagnetic radiation. Ionization can take place only if the photon carries at least the energy corresponding to the ionization potential of the atom, that is, the minimum energy required to overcome the force binding the electron within the atom.

Photon A discrete quantity of electromagnetic energy; a quantum of light. This packet of electromagnetic energy behaves like a chargeless particle travelling at the speed of light.

Photosphere The visible portion of the Sun. About 500 kilometers thick, the photosphere is a zone where the gaseous layers change from being completely opaque to radiation to being transparent. It is the layer from which the light we actually see is emitted. The temperature of the Sun's photosphere is 5780 degrees Kelvin.

Plage From the French word for "beach", a bright, dense region in the chromosphere found above sunspots or other active areas of the solar surface. They appear much brighter in hydrogen alpha than the surrounding parts of the chromosphere.

Plasma A completely ionized gas, the so-called fourth state of matter (besides solid, liquid, and gas) in which the temperature is too high for atoms as such to exist and which consists of free electrons and free atomic nuclei.

Polarization A phenomenon whereby radiation displays a given orientation of its plane of wave oscillation.

Positron A positively charged anti-particle of the electron. A subatomic particle similar in mass to an electron but carrying a positive electric charge.

Primary cosmic rays The cosmic rays that arrive at Earth's upper atmosphere from outer space.

Prominence A region of cool (100 000 degrees Kelvin), high-density gas embedded in the hot (a million degrees Kelvin), low-density solar corona; they are apparently suspended there by magnetic fields. A prominence is a filament viewed on the limb of the Sun in the light of hydrogen alpha. Quiescent prominences occur away from active regions and are stable with lifetimes of many months. They may extend upwards for tens of thousands of kilometers. Active prominences are associated with sunspots and flares. They appear as surges, sprays and loops, have violent motions, fast changes and lifetimes of up to a few hours. Eruptive prominences may arch a million kilometers outward before bursting apart; they are often associated with coronal mass ejections. *See* coronal mass ejection, filament and hydrogen alpha.

Proton A positively charged elementary particle; the nucleus of a hydrogen atom.

Proton-Proton chain (p-p chain) A series of thermonuclear reactions in which hydrogen nuclei are transformed into helium nuclei. It is the main source of energy in the Sun.

Quiet Sun The Sun when it is at the minimum level of activity in the solar cycle.

Radial velocity The velocity of an object relative to an observer along the line of sight; it is measured by the Doppler effect. To determine an object's true velocity in space, it is necessary also to know the transverse velocity, which is across the line of sight. *See* Doppler shift.

Radiation *See* Electromagnetic radiation.

Radiation belt A ring-shaped region around a planet in which electrically charged particles (usually electrons and protons) are trapped, following spiral trajectories around the direction of the magnetic field of the planet. The radiation belts surrounding the Earth are known as the Van Allen belts. Similar regions exist around other planets with magnetic fields, such as Jupiter. *See* Van Allen belts.

Radiative zone An interior layer of the Sun, lying between the core and the convection zone, where energy travels outward by radiation.

Radioactivity The spontaneous decay of certain rare, unstable, heavy nuclei into more stable light nuclei, with the release of energy.

Radioheliograph A radio telescope designed for mapping the distribution of radio emission from the Sun.

Radio radiation The part of the electromagnetic spectrum whose radiation has the longest wavelengths and smallest frequencies of all types.

Recombination The capture of an electron by a positive ion. It is the inverse process to ionization.

Redshift The increase in wavelength of electromagnetic radiation caused by the Doppler effect, when the source of radiation is moving away from the observer along the line of sight. *See* Doppler effect, Doppler shift and radial velocity.

Second of arc *See* "arc second".

Secondary cosmic rays Subatomic particles produced by collisions between primary cosmic rays and the molecules in Earth's atmosphere.

Seeing The effect of random turbulent motion in the atmosphere on the quality of the image of an astronomical object. In conditions of good seeing, images are sharp and steady; in poor seeing, they are extended and blurred and appear to be in constant motion.

Shock wave A sudden discontinuity in density and pressure propagating in a gas at supersonic velocity.

Skylab An American space station, launched into Earth orbit in May 1973. Three crews, each of three men, were sent to the station for periods of several weeks between 1973 and 1974. The station burnt up on re-entering the atmosphere in 1979.

Solar activity cycle The 11-year period between maxima (or minima) of solar activity. The solar cycle may be maintained by a dynamo driven by differential rotation and convection. Activity cycles similar to the solar cycle are apparently typical of stars with convection zones.

Solar and Heliospheric Observatory (SOHO) A satellite planned by the European Space Agency and NASA that will orbit the Sun at the point where the gravitational forces of the Earth and Sun are equal. It will carry twelve instruments designed to investigate the solar atmosphere and how it is heated, solar oscillations, how the Sun expels material into space, the structure of the Sun and processes operating within it.

Solar burst *See* burst.

Solar constant The total amount of solar energy, integrated over all wavelengths, received per unit time and unit area at the mean Sun-Earth distance outside the Earth's atmosphere. Its value is $1.37 \pm 0.02 \times 10^6$ ergs cm^{-2} s^{-1}.

Solar core The region at the center of the Sun where nuclear reactions release vast quantities of energy.

Solar corona *See* corona.

Solar eclipse A blockage of light from the Sun when the Moon is positioned precisely between the Sun and the Earth.

Solar flare *See* flare.

Solar maximum The peak of a sunspot cycle when the number of sunspots is greatest, and the output of particles and radiation is maximized.

Solar Maximum Mission (SMM) A NASA satellite launched in February 1980 for studying the Sun during a period of maximum solar activity. It failed after nine months, but repairs were successfully done by a Space Shuttle crew in 1984, and the satellite was redeployed. It re-entered the Earth's atmosphere in 1989.

Solar minimum The beginning or end of a sunspot cycle, marked by the near absence of sunspots, and a relatively low output of radiation and energetic particles.

Solar neutrino problem The discrepancy between the observed and calculated flux of neutrinos from the Sun; massive underground detectors always observe fewer neutrinos than theory says they should detect. The solar neutrino problem might be resolved if some neutrinos change form when they travel out from the Sun.

Solar sectors Regions in the solar wind that have predominantly one magnetic polarity.

Solar wind A stream of particles, primarily protons and electrons, flowing outward from the Sun at up to 900 kilometers per second. The solar wind is essentially the hot solar corona expanding into interplanetary and interstellar space.

Sound waves *See* acoustic waves.

South Atlantic Anomaly A region over the South Atlantic Ocean where the lower Van Allen belt of energetic electrically charged particles is particularly close to the Earth's surface, presenting a hazard for artificial satellites.

Spectral line A radiative feature observed in emission (bright) or absorption (dark) at a specific frequency or wavelength. It is produced by atoms as they absorb or emit light, and can be used to determine the chemical ingredients of the radiating source. Spectral lines are also used to infer the radial velocity and magnetic field of the radiator. *See* absorption line, Doppler effect, emission line, radial velocity and Zeeman effect.

Spectrograph An instrument that spreads light or other electromagnetic radiation into its component wavelengths, collectively known as a spectrum, and records the result photographically or electronically. When the instrument lacks such a recording capability, it is called a spectroscope.

Spectroheliogram A monochromatic image of the Sun produced by means of a spectroheliograph or by the use of a narrow-band filter.

Spectroheliograph An instrument for obtaining images of all or part of the Sun in monochromatic light.

Spectroscopy The study of spectra, including the position and intensity of emission and absorption lines, to determine the chemical elements or physical processes that created them.

Spectrum Electromagnetic radiation, arranged in order of wavelength from longwave radio emissions to short-wave gamma rays; also, a narrower band of wavelengths, called the visible spectrum, as when light dispersed by a prism shows its component colors. Spectra are often punctuated with emission or absorption lines, which can be examined to reveal the composition and motion of the radiating source.

Spicules Predominantly vertical, spike-like structures extending from the solar chromosphere into the corona, observed in hydrogen alpha beyond the limb. They change rapidly, having a lifetime of five to ten minutes. Typically, spicules are 1000 kilometers across and 10 000 kilometers long. They are not distributed uniformly on the Sun but concentrated along the cell boundaries of the supergranulation pattern.

Spörer minimum A period of low sunspot activity in the fifteenth century (about A. D. 1420–1570).

Stratosphere The region of the Earth's atmosphere immediately above the troposphere. It lies between heights of about 15 and 50 kilometers.

Sun The central star of the solar system. It is a dwarf star of spectral type G2 with a surface temperature of 5780 degrees Kelvin. It is a ball of hot gas and its energy source is nuclear fusion taking place in the center. Overlying this core is the radiative zone where the high-energy photons produced in the fusion reactions collide with electrons and ions to be reradiated in the form of light and heat. Beyond the radiative zone is a convection zone in which currents of gas flow upwards to release energy at the surface before flowing downwards to be reheated. The surface layer, or photosphere, from which the light we see comes, is some hundreds of kilometers thick. The layer immediately above the photosphere is the chromosphere. The tenuous, million-degree outermost layers, forming the solar corona, expand outward to form the solar wind.

Sunspot A temporary disturbed area in the solar photosphere that appears dark because it is cooler than the surrounding areas. In the dark central part of the sunspot, the umbra, the temperature is about 3700 degrees Kelvin compared with 5780 degrees Kelvin in the surrounding photosphere. Sunspots are concentrations of strong magnetic flux (2000 to 3000 Gauss), with diameters less than about 50 000 kilometers and lifetimes of a few weeks. They usually occur in pairs or in groups of opposite polarity, and move in unison across the face of the Sun as it rotates. The leading (or preceding) spot is called the P spot; the following, the F spot.

Sunspot cycle The recurring, eleven-year rise and fall in the number of sunspots. At the commencement of a new cycle sunspots erupt around latitudes of 35 to 45 degrees north and south; over the course of the cycle, subsequent spots appear closer to the equator, finishing at around 7 degrees north and south. This pattern can be demonstrated graphically as a butterfly diagram. *See* butterfly diagram and solar activity cycle.

Supergranulation cells Convective cells (about 30 000 kilometers in diameter) in the solar photosphere, distributed fairly uniformly over the solar disk, that last about 20 hours. The dominant flow in the observed cells is horizontal. The cells are outlined by the chromospheric network – boundaries containing concentrations of magnetic flux tubes. Spicules are found along the network bounding the supergranulation cells.

Surges Sudden high-velocity upwellings of radiating material along magnetic field lines in loop and active prominences, seen prominently in the light of hydrogen alpha.

Synchronous orbit *See* geosynchronous orbit.

Synchrotron radiation Electromagnetic radiation emitted by an electron travelling almost at the speed of light in the presence of a magnetic field. The name arises because it was first observed in synchrotron accelerators used by nuclear physicists. The acceleration of the electrons causes them to emit radiation. Such radiation is strongly polarized and increases in intensity at longer wavelengths. The wavelength region in which the emission occurs depends on the energy of the electron – 1 MeV electrons radiate mostly in the radio region.

Temperature A measure of the heat of an object, namely of the average kinetic energy of the randomly moving particles in an object.

Thermal bremsstrahlung Emission of radiation by energetic electrons accelerated in the field of a positive ion. *See* bremsstrahlung.

Thermal energy Energy associated with the motions of the molecules, atoms, or ions.

Thermal radiation Electromagnetic radiation arising by virtue of an object's heat (i.e. temperature), as opposed to nonthermal radiation, which is emitted by energetic electrons that are not necessarily in thermodynamic equilibrium. *See* black body radiation.

Troposphere Lowest level of the Earth's atmosphere, from the ground to about 15 kilometers above the surface. This is the region where most weather occurs.

Ultraviolet radiation The part of the electromagnetic spectrum whose radiation has frequencies somewhat greater, and wavelengths somewhat less, than those of visible light. Because the Earth's atmosphere absorbs most ultraviolet emission, thorough studies of the Sun's ultraviolet output must be conducted from space.

Ulysses A European Space Agency mission, launched on 6 October 1990, to study the interplanetary medium and the solar wind at different solar latitudes. It provides the first opportunity for measurements to be made over the poles of the Sun. Its trajectory uses the gravity assist technique to take it out of the plane of the solar system. After an encounter with Jupiter in February 1992, the spacecraft moved back towards the Sun to pass over the solar south pole in 1994 and the north pole in 1995.

Umbra The dark central region of a sunspot, where the magnetic field is vertical and typically has a strength of a few thousand Gauss. *See* penumbra.

Van Allen belts Two ring-shaped regions, that girdle the Earth's equator, in which electrically charged particles are trapped by the Earth's magnetic field. They were discovered by the United States' first successful artificial Earth satellite, Explorer 1, which was launched on 31 January 1958. The inner belt lies between 1.2 and 4.5 Earth radii (measured from the Earth's center) and the outer belt is located between 2.5 and 7.0 Earth radii. Because the Earth's magnetic field is offset from the planet's rotation axis, the inner belt dips down towards the surface in the region of the South Atlantic Ocean, off the coast of Brazil. *See* South Atlantic Anomaly.

Velocity of light The fastest speed that anything can move, equal to some 300 000 kilometers per second.

Visible spectrum The narrow range of wavelengths in the electromagnetic spectrum to which the human eye is sensitive; namely, light.

Voyager 1 and 2 Two almost identical planetary probes launched by the United States in 1977.

Wavelength The distance from crest to crest or trough to trough of an electromagnetic or other wave. Wavelengths are related to frequency. The longer the wavelength, the lower the frequency.

White light The visible portion of sunlight that includes all of its colors.

X-ray The part of the electromagnetic spectrum whose radiation has somewhat greater frequencies and smaller wavelengths than those of ultraviolet radiation. Because X-rays are absorbed by the Earth's atmosphere, X-ray astronomy is performed in space.

Yohkoh A solar satellite launched by the Japanese Institute of Space and Astronautical Science (ISAS) in August 1991. The primary scientific objective of the Yohkoh, or "sunbeam", mission is the study of high energy solar phenomena, particularly in flares, but its Soft X-ray Telescope is sufficiently sensitive for detailed observations of the ever-changing solar corona.

Zeeman effect A broadening or splitting of spectral lines caused by the influence of a magnetic field. A multiplet of lines is produced, with distinct polarization characteristics. The Zeeman effect is used to determine the direction, distribution and strength of the longitudinal magnetic fields in the photosphere. *See* magnetograph.

Quotation References

[1] An incantation from Ptolemaic Egypt. Quoted by Carl Sagan in Cosmos. Random House, New York 1980, p. 217.

[2] Nietzsche, F.: Thus Spoke Zarathustra – Of Immaculate Perception. Translated by R. J. Hollingdale. Penguin Books, Harmondsworth 1961, p. 146.

[3] Bourdillon, F. W.: Among the Flowers (1878). In: The Oxford Dictionary of Quotations, Fourth Edition (Angela Partington, ed.). Oxford University Press, New York 1992, p. 138. Also in John Bartlett's Familiar Quotations, Sixteenth Edition (Justin Kaplan, ed.). Little, Brown and Company, Boston 1992, p. 563.

[4] Letter written by Robert Bunsen to the English chemist H. E. Roscoe in November 1859. Quoted by Roscoe in: The Life and Experiences of Sir Henry Enfield Roscoe, London 1906, p. 71. It is reproduced by A. J. Meadows in his article: The Origins of Astrophysics, found in: The General History of Astronomy, Vol. 4, Astrophysics and Twentieth-Century Astronomy to 1950, Part A (Owen Gingerich, ed.), Cambridge University Press, New York 1984, p. 5. Also see G. Kirchhoff: On the Chemical Analysis of the Solar Atmosphere, Philosophical Magazine and Journal of Science 21, 185–188 (1861). Reproduced by A. J. Meadows in: Early Solar Physics, Pergamon Press, Oxford 1970, pp. 103–106; G. Kirchhoff and R. Bunsen: Chemical Analysis of Spectrum – Observations, Philosophical Magazine and Journal of Science 20, 89–109 (1860), 22, 329–349, 498–510 (1861).

[5] Thomson, W.: On the Age of the Sun's Heat, Macmillan's Magazine, March 5, 288–293 (1862). Popular Lectures I, 349–368. William Thomson is better known today as Lord Kelvin. Also see J. D. Burchfield: Lord Kelvin and the Age of the Earth. Science History Publications, New York 1975.

[6] Eddington, A. S.: The internal constitution of the stars. Nature 106, 14–20 (1920), Observatory 43, 341–358 (1920). Reproduced by K. R. Lang and O. Gingerich in: A Source Book in Astronomy and Astrophysics, 1900–1975. Harvard University Press, Cambridge, MA 1979, pp. 281–290.

[7] Eddington, A. S.: *ibid.,* reference 6.

[8] Perrin, J.: Atomes et lumière. La Revue du Mois 21, 113–166 (1920).

[9] Eddington, A. S.: The Internal Constitution of the Stars. Dover, New York 1959, p. 301 (first edition 1926).

[10] Melville, H.: Moby Dick, Marshall Cavendish Paperworks, London 1987; a reproduction of the 1922 edition, p. 141. According to Captain Ahab, all visible objects are but pasteboard masks, and man must strike through the mask. He must break through the prisoning wall. To Captain Ahab, the white whale was that wall. Nowadays, I am reminded of the song by the rock group *The Doors* entitled: Break on Through to the Other Side.

[11] Dirac, P. A. M.: Quantised Singularities in the Electromagnetic Field. Proceedings of the Royal Society of London A 133, 60–72 (1931). This paper makes the first mention of the anti-electron. Carl D. Anderson discovered it in cosmic-ray cloud chamber tracks in 1932, and in the following year proposed the name *positron*. [Anderson, C. D.: The Positive Electron. Physical Review 43, 491–494 (1933)]. However, Dirac's theory apparently played no part whatsoever in the discovery of the positron. Also see H. Kragh: Dirac – A Scientific Biography. Cambridge University Press, New York 1990.

[12] Madonna: Material Girl (1985). Written by Peter Brown and Robert Rans, published by Candy Castle Music, BMI.

[13] Updike, J.: Cosmic Gall. Originally published in: The New Yorker, 17 December 1960, p. 36. Reproduced in J. Updike: Telephone Poles and Other Poems. Alfred A. Knopf, New York 1979, p. 5, and J. Updike: Collected Poems 1953–1993. Alfred A. Knopf, New York 1993, p. 315.

[14] Pauli, W.: Remarks at the Seventh Solvay Conference, October 1933. Reproduced in the original French in: Collected Scientific Papers of Wolfgang Pauli, Vol. 2 (R. Kronig and V. F. Weisskopf, eds.). Wiley Interscience, New York 1964, p. 1319. Quoted in English by C. Sutton in: Spaceship Neutrino. Cambridge University Press, New York 1992, p. 19.

[15] Reines, F. and Cowan, C.: Telegram to Pauli dated 14 June 1956. Quoted in: Proceedings of the International Colloquium on the History of Particle Physics. Journal de Physique 43, suppl. C8, 237 (1982). Also quoted by C. Sutton in: Spaceship Neutrino. Cambridge University Press, New York 1992, p. 44.

[16] Totsuka, Y.: Recent results on solar neutrinos from Kamiokande. Nuclear Physics B19, 69–76 (1991).

[17] Bahcall, J. N. and Bethe, H. A.: A solution to the solar neutrino problem. Physical Review Letters 65, 2233–2235 (1990).

[18] Bahcall, J. N. and Bethe, H. A.: *ibid.,* reference 17.

[19] Leighton, R. B. in: Aerodynamic phenomena in stellar atmospheres – International Astronomical Union Symposium No. 12. Supplemento del Nuovo Cimento 22, 321 (1961).

[20] Leighton, R. B., Noyes, R. W. and Simon, G. W.: Velocity fields in the solar atmosphere. Preliminary report. Astrophysical Journal 135, 497 (1962).

[21] Frazier, E. N.: A spatio-temporal analysis of velocity fields in the solar photosphere. Zeitschrift für Astrophysik 68, 345 (1968).

[22] Hale, G. E.: The Earth and Sun as Magnets. Smithsonian Report for 1913, pp. 145–158. Address delivered at the Semicentennial of the National Academy of Sciences at Washington, D. C., May 1913. Reproduced by A. J. Meadows in: Early Solar Physics. Pergamon Press, Oxford 1970, pp. 291–308. Also see G. E. Hale: On the probable existence of a magnetic field in sun-spots. Astrophysical Journal 28, 315–343 (1908). Reproduced by K. R. Lang and O. Gingerich in: Source Book in Astronomy and Astrophysics, 1900–1975. Harvard University Press, Cambridge, MA 1975, pp. 96–103.

[23] Schwabe, H.: Sonnen-Beobachtungen im Jahre 1843. Astronomische Nachrichten 20, No. 495, 233–236 (1844). English translation: Solar Observations During 1843 in A. J. Meadows: Early Solar Physics. Pergamon Press, Oxford 1970, pp. 95–97.

[24] Johnson, M.: Address Delivered by the President, M. J. Johnson, Esq., on presenting the medal of the society to M. Schwabe. Monthly Notices of the Royal Astronomical Society 16, 129 (1857). Reproduced in part by H. H. Turner in: Astronomical Discovery. Edward Arnold, London 1904, pp. 156–176.

[25] Babcock, H. W.: The topology of the Sun's magnetic field and the 22-year cycle. Astrophysical Journal 133, 572 (1961).

[26] Bailey, F.: Some remarks on the total eclipse of the Sun, on July 8th 1842. Monthly Notices of the Royal Astronomical Society 5, 208–214 (1842).

[27] Biermann, L. F.: Solar corpuscular radiation and the interplanetary gas. Observatory 77, 109–110 (1957). Reproduced by K. R. Lang and O. Gingerich in: Source Book in Astronomy and Astrophysics, 1900–1975. Harvard University Press, Cambridge, MA 1979, p. 148. Also see E. N. Parker: Dynamics of the interplanetary gas and magnetic fields. Astrophysical Journal 128, 664–676 (1958).

[28] Alighieri, D.: The Comedy of Dante Alighieri, Cantica I Hell (L'Inferno). Translated by D. L. Sayers in: Basic Books, New York 1948, Canto XXVI, p. 236.

[29] Bellay, J. du: Les Regrets (1559). Quoted in John Bartlett's Familiar Quotations, Sixteenth Edition (Justin Kaplan, ed.). Little, Brown and Co., Boston 1992, p. 144.

[30] Carrington, R. C.: Description of a singular appearance seen in the Sun on September 1, 1859. Monthly Notices of the Royal Astronomical Society 20, 13–15 (1860). Richard Hodgson's account "On a curious appearance seen in the Sun" is given on the following pages 15–16. Both articles are reproduced by A. J. Meadows in: Early Solar Physics. Pergamon Press, Oxford 1970, pp. 181–185.

[31] Carrington, R. C.: ibid., reference 30.

[32] Birkeland, K.: Sur les rayons cathodiques sons l'action de forces magnétiques. Archives des Sciences Physiques et Naturelles 1, 497 (1896). Birkeland's analysis of simultaneous magnetic data at several northern stations led him to the conclusion that large currents flowed along the magnetic field lines during auroras; see K. Birkeland: The Norwegian Aurora Polaris Expedition 1902–1903. Vol. 1: On the Cause of Magnetic Storms and the Origin of Terrestrial Magnetism. H. Aschehoug Co., Christiania 1908, 1913.

[33] Van Allen, J. A., McIlwain, C. E., and Ludwig, G. H.: Radiation observations with satellite 1958 epsilon. Journal of Geophysical Research 64, 271–280 (1959). Reproduced by K. R. Lang and O. Gingerich in: Source Book in Astronomy and Astrophysics, 1900–1975. Harvard University Press, Cambridge, MA 1975, pp. 150–151.

[34] Urey, H. C.: The moon's surface features. Observatory 76, 232–234 (1956).

[35] Quoted by Robert H. Eather in: Majestic Lights – The Aurora in Science, History, and the Arts. American Geophysical Union, Washington, D.C. 1980, p. 42. From Kongespeilet (The King's Mirror) translated by L. M. Larson, New York, Twayne Publishing 1917.

[36] Quoted by Robert H. Eather in: Majestic Lights – The Aurora in Science, History, and the Arts. American Geophysical Union, Washington, D.C. 1980, p. 205. From Fridtjof Nansen: The Fram Expedition – Nansen in the Frozen World. A. G. Holman, Philadelphia, 1897.

[37] Quoted in: The Home Planet (Kevin W. Kelly, ed.), with photograph number 31. Addison–Wesley, New York 1988.

[38] Sabine, E.: Letter to John Herschel 16 March 1852. Herschel Letters No. 15.235 (Royal Society). Quoted by A. J. Meadows and J. E. Kennedy in: The origin of solar-terrestrial studies. Vistas in Astronomy 25, 420 (1982).

[39] Maunder, E. W.: Magnetic disturbances, 1882 to 1903, as recorded at the Royal Observatory, Greenwich, and their association with sunspots. Monthly Notices of the Royal Astronomical Society 65, 31 (1905). For the M regions also see J. Bartels: Terrestrial magnetic activity and its relation to solar phenomena. Journal of Geophysical Research 37, 1 (1932), and: Solar eruptions and their ionospheric effects – A classical observation and its new interpretation. Journal of Geophysical Research 42, 235 (1937). The role of coronal mass ejections in creating intense geomagnetic storms is discussed in J. Gosling: The solar flare myth. Journal of Geophysical Research 98, 18937–18949 (1993).

[40] Hoyle, F.: Lecture in 1948. Quoted by Jon Darius in: Beyond Vision. Oxford University Press, New York 1984, p. 142.

[41] Buchli, J.: Quoted in the narration for: The Blue Planet (1990).

[42] Tyndall, J.: On the absorption and radiation of heat by gases and vapours, and on the physical connexion of radiation, absorption, and conduction. Philosophical Magazine and Journal of Science 22 A, 276–277 (1861).

[43] Arrhenius, S.: On the influence of carbonic acid in the air upon the temperature of the ground. Philosphical Magazine and Journal of Science 41, 268 (1896). Arrhenius' ideas were taken up in greater detail by T. C. Chamberlin: An attempt to frame a working hypothesis of the cause of the glacial periods on an atmospheric basis. The Journal of Geology 7, 545–584 (1899). Here Chamberlin drew attention to the absorption of atmospheric carbon dioxide by the ocean, the variable depletion of carbon dioxide by weathering on globally-changing land area, and to the amplification of the greenhouse effect by increased water vapor evaporated from hotter seas.

[44] Lovelock, J. E. and Margulis, L.: Atmospheric homeostasis by and for the biosphere: The Gaia hypothesis. Tellus 26, 8 (1973).

[45] Dobson, G. M. B.: Forty years' research on atmospheric ozone at Oxford: A history. Applied Optics 7, 403 (1968).

[46] Molina, M. J. and Rowland, F. S.: Stratospheric sink for chlorofluoromethanes. Chlorine atomic-atalysed destruction of ozone. Nature 249, 810 (1974).

[47] J. Updike: Ode to Evaporation. Originally published in: The New Yorker, 31 December 1984, p. 30. Reproduced in J. Updike: Facing Nature. Alfred A. Knopf, New York 1985, pp. 79–81 and J. Updike: Collected Poems 1953–1993. Alfred A. Knopf, New York 1993, pp. 193–195.

[48] Revelle, R. and Suess, H. E.: Carbon dioxide exchange between atmosphere and ocean and the question of an increase of atmospheric carbon dioxide during the past decades. Tellus 9, 19 (1957).

[49] Hansen, J.: I'm not being an alarmist about the greenhouse effect, The Washington Post, 11 February 1989. Written by Hansen in response to criticism that his congressional testimony about the greenhouse effect was alarmist and based on faulty data.

[50] Traditional negro jubilee spiritual entitled: He's Got the Whole World in his Hands.

[51] This quotation has been attributed to the French anthropologist Maurice Chavalle's article: The Death of Nature (1955) in: M. Crichton: Congo. Ballantine Books, New York 1980, pp. 215–216. Our attempts to locate the original article have not been successful, and since it is not fully referenced by Crichton it could even be fictional, but it is nevertheless a nice way of putting things.

[52] Strand, M.: A Poem for the New Year. The New York Times, 1 January 1992.

[53] Gore, A.: Earth in the Balance. Houghton Mifflen, Boston 1992, pp. 83, 92.

[54] Solzhenitsyn, A. I.: To Tame Savage Capitalism (editorial). The New York Times, 28 November 1993.

[55] Frost, R.: Fire and Ice, in: The Poetry of Robert Frost (Edward Connery Lathem, ed.). Holt, Rinehart and Winston, New York 1979, pp. 220–221.

Further Reading

Asterisk * entries are somewhat technical.

Akasofu, S.-I.: The shape of the solar corona. Sky and Telescope 88, 24–29 (1994) – November.

New research suggests that the Sun's tenuous outer atmosphere has the same structure throughout the sunspot cycle.

Appenzeller, T.: What drives the climate? Discover 13, 64–73 (1992) – November.

Even three billion years ago, it seems, carbon dioxide was setting the global thermostat.

Bahcall, J. N.: Neutrino Astrophysics. Cambridge University Press, New York 1989.*

A very important book covering the observational, theoretical and experimental aspects of neutrino astrophysics. The solar neutrino problem is discussed in comprehensive detail, with a good introductory overview.

Böhm-Vitense, E.: Chromospheres, transition regions, and coronas. Science 223, 777–784 (1984) – 24 February.*

The Sun's chromosphere is heated by shock waves, and there is comparatively little variation in the chromospheric emission of different stars. Coronal heating and emission appear to be strongly influenced by magnetic fields, leading to large differences in the coronal X-ray emission from otherwise similar stars.

Bond, G. et al.: Correlations between climate records from North Atlantic sediments and Greenland ice. Nature 365, 143–147 (1993) – 9 September.*

Oxygen isotope measurements in Greenland ice demonstrate that a series of rapid warm-cold oscillations punctuated the last glaciation; North Atlantic sediments spanning the past 90 000 years contain a series of rapid temperature oscillations closely matching those in the ice-core record.

Brune, W. H. et al.: The potential for ozone depletion in the arctic polar stratosphere. Science 252, 1260–1266 (1991) – 31 May.*

The nature of the Arctic polar stratosphere is observed to be similar in many respects to that of the Antarctic polar stratosphere, where an ozone hole has been identified. Most of the available chlorine was converted by reactions on polar stratospheric clouds throughout the Arctic polar vortex before midwinter.

Bone, N.: The Aurora, Sun-Earth Interactions. Ellis Horwood, New York 1991.

An excellent discussion of the aurora, from historical records to the space age. Sunspots, magnetic fields, flares, coronal holes, and the solar wind are all discussed.

Brekke, A. and Egeland, A.: The Northern Light, From Mythology to Space Research. Springer-Verlag, New York 1983.

This book traces the contributions of the Scandinavians to our understanding of auroras, from Norse mythological records to later work by such important scientists as Kristian Birkeland, Carl Störmer and Hannes Alfvén.

Broecker, W. S. and Denton, G. H.: What drives glacial cycles? Scientific American 262, 49–56 (1990) – January.

Massive reorganizations of the ocean-atmosphere system, the author's argue, are the key events that link cyclic changes in the Earth's orbit to the advance and retreat of ice sheets.

Charlson, R. J. and Wigley, T. M. L.: Sulfate aerosol and climate change. Scientific American 270, 48–57 (1994) – February.

Industrial emissions of sulfur form particles that may be reflecting solar radiation back into space, thereby masking the greenhouse effect over some parts of the Earth.

Christensen-Dalsgaard, J., Gough, D., and Toomre, J.: Seismology of the Sun. Science 229, 923–931 (1985) – 6 September.*

Solar oscillations have been used to probe the inside of the Sun, determining the variation of sound speed with depth, the thickness of the convection zone, the initial helium abundance of the Sun, and the internal rotation rate.

Close, F., Marten, M., and Sutton, C.: The Particle Explosion. Oxford University Press, New York 1987.

A marvelous illustrated history of all the known particles, from the electron to the W and Z, including the neutrinos.

Cox, A. N., Livingston, W. C., and Matthews, M. S. (eds.): Solar Interior and Atmosphere. The University of Arizona Press, Tucson 1991.*

A collection of 38 technical articles by acknowledged experts, including the solar interior, solar neutrinos, solar oscillations, magnetic fields, coronal heating, solar activity, and the relation of the Sun to other stars. See, for example, the article by James F. Kasting and David H. Grinspoon on "The Faint Young Sun Problem", pp. 447–462.

Dickinson, R. E. and Cicerone, R. J.: Future global warming from atmospheric trace gases. Nature 319, 109–115 (1986) – 9 January.*

Human activity this century has increased the concentrations of atmospheric trace gases, including carbon dioxide, methane, ozone, nitrous oxide, and chlorofluorcarbons; all of these greenhouse gases can elevate global surface temperatures by blocking the escape of thermal infrared radiation.

Eather, R. H.: Majestic Lights, the Aurora in Science, History, and the Arts. American Geophysical Union, Washington, D. C. 1980.

A wonderfully comprehensive discussion of auroras, including historical documents, literature, poetry and scientific papers, with complete references.

Eddy, J. A.: The case of the missing sunspots. Scientific American 236, 80–88, 92 (1977) – May. Also see J. A. Eddy: The Maunder minimum. Science 192, 1189–1202 (1976) – 18 June.

Between 1645 and 1715 there were virtually no spots on the Sun, and this period coincided with a decrease in solar activity. The prolonged absence of sunspots is coincident with a Little Ice Age, and similar inactive periods are suggested by the relative concentration of carbon-14 in tree rings during the past 700 years.

Eddy, J. A.: A New Sun, The Solar Results from Skylab. National Aeronautics and Space Administration NASA SP-402, Washington, D. C. 1979.

An illustrated account of the scientific results of Skylab observations at ultraviolet and X-ray wavelengths, including coronal holes, solar prominences, coronal mass ejections and solar flares.

Firor, J.: The Changing Atmosphere, A Global Challenge. Yale University Press, New Haven 1990.

A nice concise description of ozone depletion and global warming.

Foukal, P. V.: Magnetic loops in the Sun's atmosphere. Sky and Telescope 62, 547–550 (1981) – December.

A good, brief discussion of solar magnetic loops seen in visible light, X-rays and radio waves.

Foukal, P. V.: The variable sun. Scientific American 262, 34–41 (1990) – February.

Its steady warmth and brightness are illusory; the Sun's output of radiation and particles varies. Systematic observations are beginning to unveil the causes of these changes and their effects on the Earth.

Foukal, P. V.: Solar Astrophysics. John Wiley and Sons, New York 1990.*

Astrophysics applied to the Sun, including radiative transfer, solar spectroscopy, plasma dynamics, internal structure, energy generation, rotation, convection, oscillations, activity, magnetism, the photosphere, chromosphere and corona, the solar wind, the heliosphere, and solar variability.

Frazier, K.: Our Turbulent Sun. Prentice-Hall, Englewood Cliffs, NJ 1982.

This good popular account focuses on the irregularities and variations of the ever-changing Sun, as well as the possible impact of the Sun on the Earth's climate and weather.

Friedman, H.: Sun and Earth. Scientific American Books, W. H. Freeman and Company, New York 1986.

Historical anecdotes and excellent illustrations highlight this very readable summary of the Sun and its effect on the Earth, including the interaction of solar radiation and energetic particles with the Earth's atmosphere.

Giovanelli, R. G.: Secrets of the Sun. Cambridge University Press, New York 1984.

A brief, richly-illustrated account of the Sun written for non-specialists, focusing on the key role of the magnetic field in explaining the hot corona, solar flares and the structure of the convection zone. There is also a good historical introduction and nice discussions of both the interior and the outer atmosphere of the Sun.

Golub, L.: What heats the solar corona? Astronomy 10, 74–85 (1982) – September.

A good description of the high-temperature corona and its possible heating mechanisms, including sound waves and magnetic heating.

Gore, A.: Earth in the Balance, Ecology and the Human Spirit. Houghton Mifflin Company, Boston 1992.

A comprehensive assessment of the forces of planetary destruction, including overpopulation, deforestation, soil erosion, ozone depletion, global warming, and water pollution, written by then Senator Al Gore, who is politically involved in environmental issues.

Gribbin, J.: Hothouse Earth, the Greenhouse Effect and Gaia. Grove Weidenfeld, New York 1990.

A fine popular account of the ice ages and the consequences and prevention of global warming.

Gribbin, J.: Blinded by the Light, the Secret Life of the Sun. Harmony Books, New York 1991.

A nice popular discussion of basic solar physics centered on the theme of what makes the Sun shine, from thermonuclear reactions in its core to the solar neutrino problem and solar oscillations.

Gough, D. and Toomre, J.: Seismic observations of the solar interior. Annual Review of Astronomy and Astrophysics 29, 627–684 (1991).*

A thorough technical account of solar and stellar oscillations, including models of the evolving Sun, oscillation modes, observations, data inversion, the solar neutrino problem, rotation and subsurface flow, mode excitation and decay, solar-cycle variations and astroseismology.

Greenland Ice-Core Project (GRIP) Members: Climate instability during the last interglacial period recorded in the GRIP ice core. Nature 364, 103–207 (1993) – 15 July.*

Isotope and chemical analyses of the GRIP ice core from Summit, central Greenland, reveal that climate in Greenland during the last interglacial period may have been characterized by a series of severe cold periods, which began rapidly and lasted from decades to centuries. If confirmed, its unstable climate raises questions about the effects of future global warming.

Hamill, P. and Toon, O. B.: Polar stratospheric clouds and the ozone hole. Physics Today 44, 34–42 (1991) – December.*

Clouds of frozen nitric acid particles that form in the polar winter stratosphere are a crucial element in the massive springtime ozone depletion over Antarctica.

Hansen, J. E. and Lacis, A. A.: Sun and dust versus greenhouse gases – An assessment of their relative roles in global climate change. Nature 346, 713–716 (1990) – 23 August.*

Many mechanisms, including variations in solar radiation and atmospheric aerosol concentrations, compete with anthropogenic greenhouse gases as causes of global climate change.

Harvey, J. W., Kennedy, J. R., and Leibacher, J. W.: GONG: To see inside our Sun. Sky and Telescope 74, 470–476 (1987) – November.

An excellent popular account of helioseismology, including recent results and the future global GONG project.

Hill, T. W. and Dessler, A. J.: Plasma motions in planetary magnetospheres. Science 252, 410–415 (1991) – April.*

Interplanetary space is pervaded by a supersonic solar wind plasma; six planets, including Earth, have magnetic fields of sufficient strength to deflect this solar wind and form a comet-shaped cavity called a magnetosphere. The main place that the solar wind enters a magnetosphere is through the long magnetotail.

Hufbauer, K.: Exploring the Sun, Solar Science Since Galileo. Johns Hopkins University Press, Baltimore, MD 1991.

A historical account that emphasizes the solar wind and the variable Sun. Techniques, theories, scientific communities, and patronage are singled out for discussion.

Imbrie, J. and Imbrie, K. P.: Ice Ages, Solving the Mystery. Enslow Publishers, Short Hills, NJ 1979.

An excellent scientific and historical discussion of the ice ages, including the periodic variations in sunlight distribution that cause them.

Kundu, M. R. and Lang, K. R.: The Sun and nearby stars – Microwave observations at high resolution. Science 228, 9–15 (1985) – 5 April.

High-resolution microwave observations are providing new insights into the nature of active regions and eruptions on the Sun and nearby stars. Magnetic changes that precede solar eruptions on time scales of tens of minutes involve primarily emerging coronal loops and the interactions of two or more loops.

Lang, K. R. and Gingerich, O.: Source Book in Astronomy and Astrophysics, 1900–1975. Harvard University Press, Cambridge, MA 1979.

A history of twentieth-century astronomy, including seminal articles on the source of solar energy, the million-degree corona, and solar magnetism.

Lean, J.: Variations in the Sun's radiative output. Reviews of Geophysics 29, 505–535 (1991) – November.*

Satellite instruments have measured, simultaneously, both the Sun's spectrally integrated radiative output and its ultraviolet spectrum. These data have been analyzed in terms of their relationships to ground-based observations that characterize different aspects of the Sun's 11-year activity cycle, allowing estimates of solar radiative output variations over time scales from days to decades.

Leibacher, J. W., Noyes, R. W., Toomre, J., and Ulrich, R. K.: Helioseismology. Scientific American 251, 48–57 (1985) – September.

Acoustic waves within the Sun are visible as oscillations on the solar surface. Their pattern and period hold clues to structure, composition and dynamics in the Sun's interior.

Meadows, A. J.: Early Solar Physics. Pergamon Press, New York 1970.

A scientific history of solar physics from 1850 to the early twentieth century, including the reproduction of several important articles.

Mitton, S.: Daytime Star, The Story of Our Sun. Faber and Faber, Boston 1981.

A very enjoyable narrative summarizes the Sun's internal properties, violent activity, relation to other stars, and solar-terrestrial relations.

National Research Council: Solar Influences on Global Change. National Academy Press, Washington, D. C. 1994.*

Several experts describe solar influences on the Earth, including solar variations and climate change, ozone variation, the middle and upper atmosphere, global warming and the Earth's near-space environment. Understanding the variable Sun and research strategies are also included.

Nicolson, I.: The Sun. Mitchell Beazley Publishers, London 1982.

A well-illustrated atlas, discussing active regions, coronal holes, filaments, flares, prominences and solar radio and X-ray radiation.

Noyes, R. W.: The Sun, Our Star. Harvard University Press, Cambridge, MA 1982.

A popular explanation of our recent knowledge of the Sun written for the educated layman.

Parker, E. N.: Magnetic fields in the cosmos. Scientific American 249, 4–45 (1983) – August.

The dynamo mechanism for the generation of magnetic fields explains why Venus has no field, why the Sun has an oscillating field, and why the dominant galactic field is parallel to the plane of the galactic disk.

Pecker, J.-C. and Runcorn, S. K. (eds.): The Earth's Climate and Variability of the Sun Over Recent Millennia – Geophysical, Astronomical and Archaeological Aspects. Philosophical Transactions of the Royal Society of London A 330, 395–687 (1990).*

A collection of 32 technical articles about the Sun's variability and its effect on the Earth during the past millennia, including carbon-14 recorded in trees, polar ice cores, Sun-weather relations, solar-cycle effects, a general discussion, and introductory and closing remarks by the editors.

Phillips, K. J. H.: Guide to the Sun. Cambridge University Press, New York 1992.

This introduction to the Sun includes the solar interior, photosphere, chromosphere and corona, the active Sun, the heliosphere, the Sun and other stars, solar energy and observing the Sun.

Radick, R. R., Lockwood, G. W., and Baliunas, S. L.: Stellar activity and brightness variations – A glimpse of the Sun's history. Science 247, 39–44 (1990) – 4 January.*

Older stars similar to the Sun tend to become brighter as their magnetic activity level increases, just as the Sun does during its 11-year activity cycle. Younger stars, however, tend to become fainter as their magnetic activity level increases.

Ramanathan, V.: The greenhouse theory of climate change. Science 240, 293–299 (1988) – 15 April.*

Since the dawn of the industrial era, the atmospheric concentrations of several radiatively active gases have been increasing as a result of human activities. The radiative heating from this inadvertent experiment has driven the climate system out of equilibrium with the incoming solar energy. The predicted changes, during the next few decades, could far exceed natural climate variations in historical times.

Ruddiman, W. F. and Kutzbach, J. E.: Plateau uplift and climatic change. Scientific American 264, 66–75 (1991) – March.

The formation of giant plateaus in Tibet and the American West may explain why Earth's climate has grown markedly cooler and more regionally diverse in the past 40 million years.

Sackmann, I.-J., Boothroyd, A. I., and Kraemer, K. E.: Our Sun – Present and future. Astrophysical Journal 418, 457–468 (1993).*

Evolutionary models are used to compute the Sun's future, including its increased brightness and expansion to a giant star, with a resultant runaway greenhouse catastrophe on the Earth.

Sagdeev, R. Z. and Kennel, C. F.: Collisionless shock waves. Scientific American 264, 106–113 (1991) – April.

To many theorists' surprise, shock waves form even in the rarefied material between the planets. These shocks sculpt the space environment and lie at the heart of a variety of astrophysical phenomena.

Schneider, S. H.: Global Warming, Are We Entering the Greenhouse Century? Sierra Club Books, San Francisco 1989. Also see Schneider, S. H.: The greenhouse effect – Science and policy. Science 243, 771–781 (1989) – 10 February.

A good account of a threat of future global warming caused by human activity, including model predictions and political considerations.

Smithsonian Institution: Fire of Life. W. W. Norton and Company, New York 1981.

A lavish, colorful book that considers the Sun from every conceivable aspect – as an object of ancient worship, the hub of the planets, the source and pacemaker of life on Earth, a typical evolving star in our Galaxy, and a future source of energy through photovoltaic cells on the ground and in space.

Sonnett, C. P., Giampapa, M. S., and Matthews, M. S. (eds.): The Sun in Time. The University of Arizona Press, Tucson 1991.*

A collection of 36 technical articles on energetic solar particles, solar isotopes, the Sun and climate, the early Sun and solar-like stars.

Stahl, P. A.: Prominences. Astronomy 11, 66–71 (1983) – January.

A general account that includes the magnetic support of prominences, and the condensation or injection of prominence material.

Sturrock, P. A., Holzer, T. E., Mihalas, D. M., and Ulrich R. K. (eds.): Physics of the Sun. D. Reidel, Boston 1986.*

A comprehensive three-volume collection of technical articles on the solar interior, the solar atmosphere, solar astrophysics and solar-terrestrial relations.

Sutton, C.: Spaceship Neutrino. Cambridge University Press, New York 1992.

A really fine historical and scientific discussion, on the popular level, of the prediction, nature, and discovery of neutrinos, including neutrinos from the Sun and the pivotal role the neutrino has played in modern particle physics.

Tans, P. P., Fung, I. Y., and Takahashi, T.: Observational constraints on the global atmospheric carbon dioxide budget. Science 247, 1431–1438 (1990) – 23 March.*

The northern oceans may not be a major sink of carbon dioxide, and a large amount of the waste gas is apparently absorbed on the continents by terrestrial ecosystems.

Taylor, P. O.: Observing the Sun. Cambridge University Press, New York 1991.

An account of how amateur astronomers might observe the Sun, including a historical background, new techniques and equipment, observations of sunspots and eclipses, and reporting observations.

Toon, O. B. and Turco, R. P.: Polar stratospheric clouds and ozone depletion. Scientific American 264, 68–74 (1991) – June.

Clouds rarely form in the dry, Antarctic stratosphere, but when they do, they chemically conspire with chlorofluorocarbons to create the ozone hole that opens up every spring.

Waters, J. W. et al.: Stratospheric chlorine monoxide and ozone from the microwave limb sounder on the upper atmosphere research satellite. Nature 362, 597–602 (1993) – 15 April.*

Chlorine in the lower stratosphere was almost completely converted to chemically reactive forms in both the northern and southern polar winter vortices.

Weneser, J. and Friedlander, G.: Solar neutrinos – Questions and hypotheses. Science 235, 755–759 (1986) – 13 February.*

The discrepancy between the observed and predicted flux of neutrinos from the Sun can be resolved by changes in the standard solar model or by alterations of neutrinos during their transit from the Sun. Also see the following article on future experiments that may decide between the two explanations.

Wentzel, D. G.: The Restless Sun. Smithsonian Institution Press, Washington, D. C. 1989.

A nonmathematical account of the internal solar energy source, magnetic activity on the Sun, solar and stellar physics, solar-terrestrial effects, and contemporary solar research.

Wolfson, R.: The active solar corona. Scientific American 248, 104–119 (1983) – February.

The solar corona is a restless, dynamic structure of ever-changing aspect. The continuous solar wind flows from coronal holes and is punctuated by eruptive outbursts from active regions. Magnetic fields mold the coronal gas into its intricate shape, and it is magnetic forces that are responsible for the detail, variety and evolution of coronal structure.

Author Index

Subject Index